Artificial Intelligence and Optimization Techniques for Smart Information System Generations

This book stands on the shoulders of sustainability engineering insights to identify blind spots in conventional business, engineering, and environmental school curricula that result in substantial missed opportunities. Harvesting these financial gains provides resources to secure environmental and social outcomes. Cutting through the clutter of terminology and approaches, this book uses case studies to outline a proven path to profitably secure "net zero" for facilities and their supply chains. And rather than stopping at the uninspiring target of "doing no harm" it further empowers businesses and other organizations to create additional economic, environmental, and social benefits.

- Offers numerous case studies to illustrate how to reduce environmental footprints while increasing profit margin and business value.
- Explores redemptive entrepreneurship through the lens of the author's founding of an award-winning engineering firm.
- Presents strategies to strengthen the curricula of engineering, business, accounting, and procurement programs by uncovering significant sources of untapped value in conventional teaching.

Future Generation Information Systems
Series editor- Bharat Bhushan

With the evolution of future generation computing systems, it becomes necessary to occasionally take stock, analyze the development of its core theoretical ideas, and adapt to radical innovations. This series will provide a platform to reflect the theoretical progress, and forge emerging theoretical avenues for the future generation information systems. The theoretical progress in the Information Systems field (IS) and the development of associated next generation theories is the need of the hour. This is because Information Technology (IT) has become increasingly infused, interconnected and intelligent in almost all context.

Industry 6.0: Technology, Practices, Challenges, and Applications
Kishor Kumar Reddy C, Srinath Doss, Lavanya Pamulaparty, Kari J Lippert and Ruchi Doshi

Convergence of Deep Learning and Artificial Intelligence in Internet of Things
Ajay Rana, Arun Rana, Sachin Dhawan, Sharad Sharma and Ahmed A. Elngar

Next Generation Communication Networks for Industrial Internet of Things Systems
Sundresan Perumal, Mujahid Tabassum, Moolchand Sharma and Saju Mohanan

Artificial Intelligence and Optimizing Techniques for Smart Information System Generations
Aleem Ali, Rajdeep Chakraborty and Nawaf R. Alharbe

Artificial Intelligence and Optimization Techniques for Smart Information System Generations

Edited by
Aleem Ali, Rajdeep Chakraborty and
Nawaf R. Alharbe

CRC Press is an imprint of the
Taylor & Francis Group, an **informa** business

First edition published 2025
by CRC Press
2385 NW Executive Center Drive, Suite 320, Boca Raton FL 33431

and by CRC Press
4 Park Square, Milton Park, Abingdon, Oxon, OX14 4RN

CRC Press is an imprint of Taylor & Francis Group, LLC

© 2025 selection and editorial matter, Aleem Ali, Rajdeep Chakraborty and Nawaf R. Alharbe individual chapters, the contributors

Reasonable efforts have been made to publish reliable data and information, but the author and publisher cannot assume responsibility for the validity of all materials or the consequences of their use. The authors and publishers have attempted to trace the copyright holders of all material reproduced in this publication and apologize to copyright holders if permission to publish in this form has not been obtained. If any copyright material has not been acknowledged please write and let us know so we may rectify in any future reprint.

Except as permitted under U.S. Copyright Law, no part of this book may be reprinted, reproduced, transmitted, or utilized in any form by any electronic, mechanical, or other means, now known or hereafter invented, including photocopying, microfilming, and recording, or in any information storage or retrieval system, without written permission from the publishers.

For permission to photocopy or use material electronically from this work, access www.copyright.com or contact the Copyright Clearance Center, Inc. (CCC), 222 Rosewood Drive, Danvers, MA 01923, 978-750-8400. For works that are not available on CCC please contact mpkbookspermissions@tandf.co.uk

Trademark notice: Product or corporate names may be trademarks or registered trademarks and are used only for identification and explanation without intent to infringe.

ISBN: 978-1-032-71703-6 (hbk)
ISBN: 978-1-032-97258-9 (pbk)
ISBN: 978-1-003-59296-9 (ebk)

DOI: 10.1201/9781003592969

Typeset in Times
by Newgen Publishing UK

Contents

About the Editors ix
About Series Editors xi
Preface..xiii

Chapter 1 Unveiling the Power of Prediction: A Comprehensive Guide to Machine Learning Techniques, from Data Preparation to Model Interpretability for Early Prediction of Diabetes and Effective Management 1

Inderdeep Kaur and Aleem Ali

Chapter 2 Digital Image Forgery Techniques for Smart Information Generations.. 24

Gurmeet Kaur Saini and Salah Al-Majeed

Chapter 3 Artificial Intelligence (AI) and Optimization for Health Information System (HIS) Generation.................................... 48

Payal Thakur, Shanu Khare, and Navjot Singh Talwandi

Chapter 4 Agricultural Information Systems Emphasizing Agro Robots Towards Digital and Sustainable Agriculture: A Scientific Review 70

P. K. Paul, Nilanjan Das, Rajibul Hossain, Mustafa Kayyali, and Ricardo Saavedra

Chapter 5 Smart Agriculture Based on Artificial Intelligence in the African Region: Open Challenges, Solutions..................................... 89

Ametovi Koffi Jacques Olivier, Taushif Anwar, Ghufran Ahmad Khan, and Zubair Ashraf

Chapter 6 Synergizing Artificial Intelligence and Optimization Techniques for Enhanced Public Services... 107

Kapil Saini, Ravi Saini, and Parul Sehrawat

Chapter 7 Architectural Pattern for Implementing XAI as a Service for Container-Orchestrated Machine Learning Model Deployments 128

Amit Chakraborty, Susmita Ganguly, and Saptarshi Das

Chapter 8 Quantum Machine Learning (QML) Algorithms for Smart Biomedical Applications 150

Inzimam Ul Hassan, Swati and Aakansha Khanna

Chapter 9 Artificial Intelligence for Information System Security 172

Anup Lal Yadav, Navjot Singh Talwandi, Shanu Khare, and Payal Thakur

Chapter 10 Optimized Image Recognition for Smart Information Generations: A Comprehensive Evaluation of Image Processing Techniques 191

Gurmeet Kaur Saini, Aleem Ali, and Nawaf R. Alharbe

Chapter 11 Machine Learning-based Security Algorithms for Cloud Computing: A Comprehensive Survey 220

Shaweta Sachdeva, Aleem Ali, and Ahmed A. Elngar

Chapter 12 Securing Information in Transit: Leveraging AI/ML for Robust Data Protection 232

Navjot Singh Talwandi and Kulvinder Singh

Chapter 13 Optimizing Medical Image Compression for Efficient Information System Integration: A Comprehensive Review 245

Kanwaldeep Kaur Sidhu, Heena Talat, and Intisar S. Al-Mejibli

Chapter 14 FaceTrack: A Face Recognition-based Real-time Attendance Marking Approach using Haar Cascade and Machine Learning 258

R. Rafeek, V. S. Anoop, and Zahid Akhtar

Chapter 15 Optimizing Information Systems for Green Computing
in Higher Education: From Awareness to Action 278

*Tanvir Chowdhury, Habibur Rahman, Monjurul
Islam Sumon, and Md Fokrul Akon*

Chapter 16 Evaluating Thin Client Solutions: A Sustainable and
Energy-Efficient Alternative to Traditional Classroom PCs 288

*Md Tanvir Chowdhury, Habibur Rahman,
Monjurul Islam Sumon, Md Mahmud Hasan Akhnoda,
and Samia Shameem Haque*

Index 299

About the Editors

Aleem Ali is presently working as a Professor in the Department of Computer Science and Engineering at Chandigarh University, Mohali, Punjab, India. He has published more than fifty research papers in journals and conferences of national and international repute. His areas of research interest include the Internet of Things, Data Mining, Machine Learning Techniques, Wireless Sensor Networks, Network Security, and Cryptography. Dr. Ali plays a pivotal role in shaping the future of computer science and engineering.

Professor
Department of Computer Science & Engineering,
University Institute of Engineering, Chandigarh University,
Punjab - 140413.
Website: https://www.cuchd.in/
E-mail: aleem08software@gmail.com

Rajdeep Chakraborty is currently working as a Professor in the Department of Computer Science and Engineering at Medi-Caps University, Indore, MP, India. He has more than twenty years of research and academic experience. He has publications in reputed international journals and conferences and has authored a book on Hardware Cryptography. His fields of interest are primarily Cryptography and Computer Security.

Professor
Department of Computer Science & Engineering,
Medi-Caps University,
Rau, Indore - 453331,
Madhya Pradesh, India.
E-mail: rajdeep.research@gmail.com

Nawaf R. Alharbe is presently working as a Professor in the College of Computer Science and Engineering, Department of AI and Data Science, at Taibah University, Madinah, Saudi Arabia. He is also an esteemed Visiting Professor in the Faculty of Computing, Engineering, and Sciences at Staffordshire University, United Kingdom. He has published research papers in journals and conferences of national and international repute. His areas of research interest include the

Internet of Things, Network Communication, Sensor Technology, Network Security, Wireless Communications, Computer Networking, and Cloud Computing.

Professor
Taibah University, Madinah,
KSA - 00966-506313085
E-mail: nrharbe@taibahu.edu.sa

About Series Editors

Dr. Bharat Bhushan is an Associate Professor of the Department of Computer Science and Engineering (CSE) at the School of Engineering and Technology, Sharda University, Greater Noida, India. For three consecutive years (2021 to 2023), Stanford University (USA) listed Dr. Bharat Bhushan in the top 2% scientists list. He received his Undergraduate Degree (B-Tech in Computer Science and Engineering) with Distinction in 2012, his Postgraduate Degree (M-Tech in Information Security) with Distinction in 2015 and his Doctorate Degree (PhD Computer Science and Engineering) in 2021 from Birla Institute of Technology, Mesra, India. He earned numerous international certifications such as CCNA, MCTS, MCITP, RHCE and CCNP. He has published more than 150 research papers in various renowned International Conferences and SCI-indexed journals including the *Journal of Network and Computer Applications* (Elsevier), *Wireless Networks* (Springer), *Wireless Personal Communications* (Springer), *Sustainable Cities and Society* (Elsevier) and *Emerging Transactions on Telecommunications* (Wiley). He has contributed to more than 50 book chapters in various books and has edited 30 books from the most famed publishers like Elsevier, Springer, Wiley, IOP Press, IGI Global, and CRC Press. He is a series editor of two prestigious Scopus Indexed Book Series named CMIA (Computational Methods for Industrial Applications) and FGIS (Future Generation Information System) published by CRC Press, Taylor and Francis, USA.

Dr. Bharat Bhushan
Associate Professor, Computer Science &
Engineering (CSE)
Email: bharat.bhushan@sharda.ac.in

Preface

In the rapidly evolving landscape of technology, the convergence of Artificial Intelligence (AI) and optimization techniques has emerged as a powerful catalyst for innovation across various domains. This book, "*Artificial Intelligence and Optimization Techniques for Smart Information System Generations*," seeks to explore this dynamic synergy, presenting a comprehensive guide to the theoretical foundations, advanced applications, and practical implementations of these technologies.

The impetus for this book stems from the growing need to understand how AI and optimization can be harnessed to create smarter, more efficient information systems. As researchers, practitioners, and educators, we have witnessed firsthand the transformative potential of these technologies. Our goal is to provide a resource that not only elucidates the core principles of AI and optimization but also demonstrates their applicability in solving real-world problems.

This book is organized into three parts, each designed to cater to a specific aspect of AI and optimization:

1. **Foundational Concepts and Applications of AI**: This section lays the groundwork by exploring fundamental AI techniques and their applications. From machine learning and digital image forgery detection to health information systems, the chapters provide a solid understanding of how AI can be leveraged to generate intelligent solutions.
2. **Advanced AI Applications and Techniques**: Building on the foundational knowledge, this section delves into more complex and innovative applications. Topics such as smart agriculture, AI-augmented health systems, and quantum machine learning are discussed, showcasing the cutting-edge advancements in the field.
3. **AI in Security and Specialized Applications**: The final section focuses on the integration of AI in security and specialized domains. It addresses critical areas such as information system security, medical image compression, and the classification of gravitational waves, highlighting the versatility and impact of AI across diverse sectors.

Throughout the book, we have included contributions from esteemed researchers and practitioners from around the world. Their insights and expertise have been instrumental in shaping the content, ensuring that it is both comprehensive and up-to-date. We are grateful for their collaboration and dedication to advancing the field of AI and optimization.

We also recognize the importance of bridging the gap between academia and industry. Therefore, this book is designed to be a valuable resource for students, researchers, and professionals alike. Whether you are seeking to deepen your understanding of AI and optimization or looking for practical solutions to implement in your work, we hope this book serves as a useful guide.

As we look to the future, the potential for AI and optimization to drive innovation and efficiency in information systems is immense. We hope this book inspires new ideas, fosters collaboration, and contributes to the ongoing development of smarter, more intelligent technologies.

We extend our heartfelt thanks to everyone who has supported this endeavor, and we look forward to the continued advancements in this exciting field.

Dr. Aleem Ali, Dr. Rajdeep Chakraborty, and Dr. Nawaf R. Alharbe
June 2024

1 Unveiling the Power of Prediction

A Comprehensive Guide to Machine Learning Techniques, from Data Preparation to Model Interpretability for Early Prediction of Diabetes and Effective Management

Inderdeep Kaur and Aleem Ali

1.1 INTRODUCTION

Diabetes is a chronic medical illness characterised by high amounts of blood glucose, sometimes known as blood sugar. This condition occurs when the body is unable to create enough insulin or to use the insulin that it does make adequately [1]. Insulin, a pancreatic hormone, is essential for controlling blood sugar levels and allowing glucose entrance into cells, where it is utilised as energy. Type 1 diabetes and Type 2 diabetes are the two most common forms. It is shown in Figure 1.1.

- *Type 1 Diabetes*: This kind of diabetes develops when the immune system assaults and kills insulin-producing beta cells in the pancreas. Type 1 diabetes patients require insulin injections to control their blood sugar levels.
- *Type 2 Diabetes*: This is the most prevalent kind of diabetes and is caused by a combination of insulin resistance (cells that do not respond adequately to insulin) and a relative insulin deficit. Sedentary behaviour, bad nutritional habits, and obesity are frequently related with it.

Diabetes prevalence has been increasing internationally, posing a huge public health problem. Diabetes affects millions of individuals globally, according to the World Health Organization (WHO), and the number is growing. Diabetes is recognised as a worldwide disease that affects both developed and developing countries. Changing

DOI: 10.1201/9781003592969-1

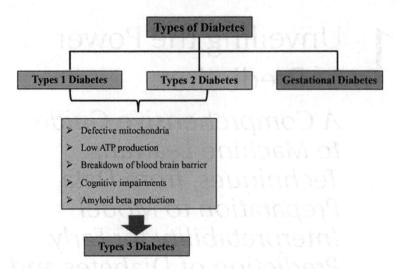

FIGURE 1.1 Types of Diabetes[3].

food habits, decreased physical exercise, and an ageing population all contribute to the rising incidence.

Diabetes prevalence varies by location and demographic group. Certain populations, such as Indigenous cultures and ethnic groupings, may be predisposed to diabetes [1]. Diabetes has traditionally been linked with older age groups, but there is a disturbing trend of diabetes appearing in younger populations, including children and adolescents. This is frequently associated with lifestyle issues such as poor nutrition and sedentary behaviour. The increased incidence of diabetes places a significant strain on healthcare systems across the world. Diabetes complications, such as cardiovascular disease and renal failure are very common [2].

1.2 IMPORTANCE OF EARLY PREDICTION IN EFFECTIVE MANAGEMENT

Diabetes prediction is critical to ensure proactive and successful management of this chronic illness. Recognising and treating diabetes early on can have far-reaching consequences for individual health outcomes, healthcare systems, and public health [3]. Here's a more in-depth look at the significance:

1. Complications Prevention:
    - Microvascular problems: Prompt management to avoid microvascular problems such as retinopathy, nephropathy, and neuropathy is possible with early prediction.
    - Macrovascular Complications: Treating diabetes early minimises the incidence of macrovascular complications such as cardiovascular disorders, for instance heart attacks and strokes.

2. Improved Quality of Life:
   - Reduced Symptom Burden: Early identification allows for the implementation of treatment methods prior to the onset of major symptoms, therefore enhancing the overall quality of life for diabetics.
   - Pancreatic Function Preservation: In some circumstances, early intervention may help maintain pancreatic function, particularly in Type 2 diabetes, where lifestyle changes might be critical.
3. Targeted and Personalised Interventions:
   - Tailored Treatment Programmes: Early prediction enables the formulation of personalised treatment programmes that take individual risk factors, lifestyle variables, and genetic predispositions into consideration.
   - Precision Medicine Methodologies: With advances in healthcare technology, early prediction makes precision medicine methods easier to apply, optimising therapy tactics for each patient.
4. Cost-Effectiveness
   - Lower Healthcare Costs: Early management is frequently more cost-effective than dealing with problems that may emerge if diabetes is left untreated or inadequately controlled [4].
   - Resource Allocation: When diabetes is detected early, healthcare resources may be used more efficiently, avoiding the need for intense and costly therapies later on.
5. Empowerment Through Education
   - Patient Education: Early prediction enables prompt education of patients at risk, arming them with knowledge about lifestyle adjustments, dietary changes, and self-management measures.
   - Behavioural therapies: When begun early, behavioural therapies like increasing physical exercise and good diet are more successful, potentially avoiding the progression of prediabetes to diabetes.
6. Public Health Impact
   - Reduced Disease Burden: A population-wide strategy for early prediction and intervention can help to reduce the total burden of diabetes on public health [5].
   - Preventive Health Campaigns: Early prediction enables the establishment of focused public health campaigns to raise awareness, promote screening, and encourage healthy behaviours.
7. Research Opportunities
   - Identification of Biomarkers: Early prediction offers up options for research into discovering new biomarkers that may improve the accuracy and specificity of predictive models.
   - Understanding Disease Development: Researching people in the early stages of diabetes can reveal vital insights into the disease's natural history and development [6].

1.3 AN OVERVIEW OF DIABETES TYPES 1 AND 2

1.3.1 Type 1 Diabetes

Autoimmune Destruction: One of the hallmarks of Type 1 diabetes is the immune system's inadvertent attack on and destruction of the pancreatic beta cells that produce insulin. This results in little or no insulin production by the pancreas, which causes a complete insulin deficit. Although it can strike at any age, Type 1 diabetes typically first appears in childhood or adolescence. Symptoms may appear suddenly or develop quickly. People who have a family history of Type 1 diabetes may be at a higher risk due to a genetic vulnerability. The autoimmune response may be triggered by viral infections or other environmental causes. To replace the hormone that is lacking in Type 1 diabetes, insulin therapy must be administered for the rest of one's life. To modify insulin dosages, blood glucose levels must be checked on a regular basis. Only a small proportion of diabetes cases—roughly 5–10% of all cases diagnosed—have Type 1 diabetes. The two types of diabetes and their common characteristics are shown in Figure 1.2.

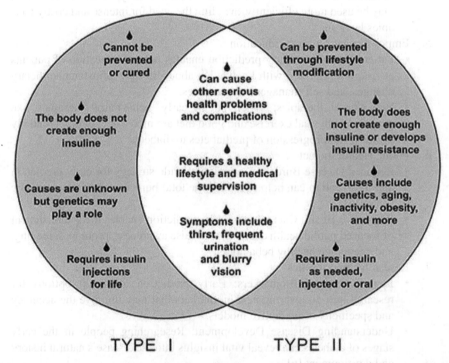

FIGURE 1.2 Type 1 and Type 2 Diabetes [12].

1.3.2 Type 2 Diabetes

A hallmark of Type 2 diabetes is the ineffectiveness of cells' response to insulin. This condition is brought on by the pancreas producing less insulin than required over time. Although there is an increasing frequency among younger groups, Type 2 diabetes usually manifests in maturity. The symptoms may not appear as severe at first because of the delayed start. Two of the biggest risk factors are obesity, especially abdominal obesity, and a sedentary lifestyle. The risk of Type 2 diabetes is influenced by genetic, ethnic, and family history variables. Patients are first advised to make lifestyle modifications to their food and exercise routines. As the condition worsens, doctors may prescribe insulin therapy as well as oral drugs. Type 2 diabetes is far more common than Type 1 diabetes globally, accounting for the majority of diabetes occurrences.

1.3.3 Common Characteristics

a) Hyperglycemia: Hyperglycemia is a result of raised blood glucose levels, which are caused by both forms of diabetes.
b) Extended hyperglycemia increases the risk of problems pertaining to the kidneys, eyes, nerves, and heart.
c) Diet and Exercise: Managing both forms of diabetes requires lifestyle changes, such as eating a balanced diet and getting regular exercise.
d) Blood Glucose Monitoring: For efficient self-management, blood glucose levels must be regularly checked.
e) Long-Term Complications: If diabetes is not well controlled, both Type 1 and Type 2 diabetes might result in long-term problems.
f) Cardiovascular Disease: People with diabetes are more likely to develop cardiovascular illnesses.

1.3.4 Distinctive Characteristics

Understanding the distinctive characteristics of Type 1 and Type 2 diabetes is essential for early diagnosis. Even though both categories have similar risk factors, knowing the unique traits makes prediction tactics more effective. The following characteristics make an early forecast unique:

Type 1 Prediction:

1. Genetic Indices: There is frequently a significant hereditary component to Type 1 diabetes, with certain genetic markers linked to heightened vulnerability [23]. Early prediction efforts can be strengthened by genetic testing, which can identify individuals with a familial tendency.
2. Autoimmune Biomarkers: In Type 1 diabetes, autoimmune markers, such as autoantibodies directed against pancreatic beta cells, signal the immune system's assault. Blood test results including autoimmune biomarkers may indicate a continuing autoimmune response even in the absence of clinical symptoms.
3. Onset Age: Children or adolescents with Type 1 diabetes often experience its symptoms. Determining the age of commencement facilitates concentrating prediction efforts on younger groups and enacting close observation.

4. Quick onset of symptoms: Characteristic: Signs of Type 1 diabetes, such as increased thirst, frequent urination, and unexplained weight loss, can manifest quickly. Quick diagnostic testing for early intervention is prompted by identifying patients with sudden onset of symptoms.
5. Antibodies against Islet cells: One of the main characteristics of Type 1 diabetes is the existence of islet cell autoantibodies, which include antibodies against insulin, GAD65, and IA-2. The presence of these antibodies in those who are at risk indicates an autoimmune reaction and an increased chance of developing diabetes.

Type 2 Prediction:

1. Insulin Resistance: Unique Aspect of Type 2 Diabetes: Cells with insulin resistance do not react to insulin as well. The likelihood of Type 2 diabetes can be predicted by early markers of insulin resistance, such as higher fasting insulin levels.
2. Obesity and Central Adiposity: One of the main risk factors for Type 2 diabetes is obesity, particularly central adiposity, or abdominal obesity. Keeping an eye on waist size and body weight helps identify those who are at risk, especially those who have accumulated.
3. Age and Family History: Adults with a family history of Type 2 diabetes are more likely to develop the condition [24]. Risk evaluations are guided by age and family history, which encourages early monitoring and preventative interventions.
4. Gradual Onset of Symptoms: Increased thirst and weariness are common Type 2 diabetes symptoms that appear gradually. Early screening for those with risk factors is prompted by the identification of modest, persistent symptoms [25].
5. Impaired Glucose Tolerance (IGT) and Impaired Fasting Glucose (IFG): Before acquiring overt diabetes, people with prediabetes may show signs of impaired glucose tolerance (IGT) or impaired fasting glucose (IFG). Early therapies to prevent or postpone the onset of Type 2 diabetes can be made possible by identifying persons with prediabetes.
6. Elements of the Metabolic Syndrome: People at risk for Type 2 diabetes frequently have metabolic syndrome symptoms, such as high blood pressure, dyslipidaemia, and raised fasting glucose. Tracking the elements of the metabolic syndrome helps identify people who have a combination of risk factors [26].

1.4 TRADITIONAL TECHNIQUES FOR DIABETES PREDICTION

The identification of those at risk of acquiring diabetes and the implementation of preventive interventions have been made possible by the use of traditional approaches for diabetes prediction. These techniques use demographic data, historical data, and clinical risk factors to determine a person's probability of acquiring diabetes [13]. Some Common techniques are shown in Figure 1.3. Although these techniques have

Machine Learning for Diabetes: From Data Prep to Early Prediction 7

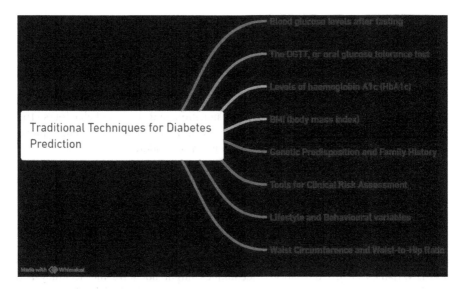

FIGURE 1.3 Traditional Techniques for Diabetes Prediction.

shown to be beneficial, machine learning techniques aim to get around some of their drawbacks. This is a thorough analysis of conventional techniques:

1. Blood glucose levels after fasting: Following a minimum of eight hours without food or liquids (except water), a person's fasting blood glucose levels are assessed [14]. High fasting blood glucose levels are a crucial component of the diabetes diagnostic criteria and are a predictor of the disease's likelihood to manifest.
2. The OGTT, or oral glucose tolerance test: The OGTT evaluates the body's capacity to metabolise glucose by taking blood glucose readings both before and after ingesting a glucose solution [15]. OGTT is a diagnostic tool for prediabetes and diabetes that provides information on how the body metabolises glucose over time.
3. Levels of haemoglobin A1c (HbA1c): HbA1c gives a longer-term view of glycemic management by reflecting the average blood glucose levels over the previous two to three months. Inadequate management of diabetes is indicated by elevated HbA1c readings. Tracking changes in HbA1c is a popular way to determine the likelihood of developing diabetes.
4. BMI (body mass index): Body mass index (BMI) is a weight-and-height-based indicator of body fat. Since a higher BMI is linked to a higher risk of Type 2 diabetes, it is frequently employed as an indicator to forecast the likelihood of developing the disease.
5. Genetic Predisposition and Family History: An individual's vulnerability to diabetes is influenced by genetic and family history variables. People who have a genetic marker for diabetes or a family history of the disease may be deemed to be at higher risk, which helps with predictive evaluations.

6. Tools for Clinical Risk Assessment: To estimate the risk of acquiring diabetes, a variety of clinical risk assessment techniques combine variables including age, gender, family history, and lifestyle. By offering a methodical approach to assessing a person's risk profile, these tools help medical professionals make recommendations for preventative treatment.
7. Lifestyle and Behavioural Variables: Diabetes risk is influenced by lifestyle variables, including physical activity, nutrition, and smoking behaviours [16]. Examining a person's behaviour and way of life might assist in finding modifiable risk factors that lifestyle interventions can address.
8. Waist Circumference and Waist-to-Hip Ratio: Insulin resistance and a higher risk of diabetes are linked to abdominal obesity as shown by measurements such as waist circumference and waist-to-hip ratio. These anthropometric measurements help determine diabetes risk and are used to detect central adiposity.

Traditional approaches have some limitations. First, they often offer a snapshot of a person's health at a particular moment in time, which may leave out dynamic changes in risk variables [17]. Conventional techniques might not have the accuracy and specificity required to correctly identify those who are at high risk. Conventional approaches frequently fail to make use of complex datasets, such as genetic and lifestyle data, which might support a more thorough risk assessment.

1.5 INTRODUCTION TO MACHINE LEARNING AND ITS POTENTIAL IN DIABETES PREDICTION

Machine learning (ML), a type of artificial intelligence, has emerged as a transformational force in healthcare, allowing for earlier prediction, diagnosis, and personalised therapy. In the context of diabetes prediction, ML algorithms use data analytics to uncover patterns, correlations, and predictive insights. This section introduces machine learning in healthcare, concentrating on its potential for diabetes prediction. Generally, machine learning refers to a computer system's capacity to learn and improve based on experience without being explicitly programmed [7]. It entails algorithms that can recognise patterns, forecast outcomes, and adapt to changing conditions. Various machine learning methodologies are supervised learning, unsupervised learning, and reinforcement learning, etc.

1.5.1 Role of Machine Learning in Healthcare

Medical data analysis may benefit greatly from the use of machine learning algorithms, which have the potential to transform medical diagnosis, treatment, and results. The role of machine learning in healthcare is shown in Figure 1.4. The following are a few prominent and well-known uses of machine learning algorithms in the analysis of medical data:

a) Disease diagnosis: Models that correctly identify a range of medical disorders may be created using machine learning methods. For instance, highly accurate machine learning models have been constructed to identify heart disease, skin

Machine Learning for Diabetes: From Data Prep to Early Prediction 9

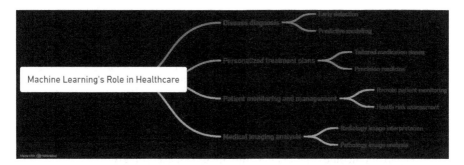

FIGURE 1.4 Machine Learning in Healthcare.

cancer, and other illnesses. These algorithms are able to swiftly and precisely diagnose patients by analysing patient data, including test findings or medical imaging.

b) Drug discovery: To find and forecast the efficacy of possible drug candidates, machine learning techniques can be employed. Machine learning algorithms can find chemicals that are likely to be useful in treating particular medical diseases by analysing enormous databases of molecular structures. This can lower expenses and expedite the drug development process significantly [9].

c) Predictive modelling: Models that forecast the chance of different medical occurrences, including hospital readmissions or the beginning of a medical disease, may be created using machine learning techniques. Healthcare professionals may use these models to identify patients who are at risk of negative outcomes and take preventative measures before the event happens.

d) Personalised medicine: Based on each patient's unique genetics, medical history, and lifestyle choices, personalised treatment programmes may be created using machine learning algorithms. Machine learning models, for instance, may be used to tailor treatment regimens to each patient's specific needs and forecast how they will respond to treatments.

e) Image analysis: Machine learning algorithms may be used to analyse medical images, including X-rays, MRIs, and CT scans. These images are rich sources of data. Medical picture anomalies, such as tumours or other lesions, may be recognised by ML models, and based on these results, precise diagnoses can be made [9].

f) Analysis of electronic health records: Machine learning algorithms may be used to examine the abundance of data found in electronic health records (EHRs). Patterns in EHR data, such as risk factors for particular medical illnesses or the effectiveness of particular therapies, can be found using machine learning algorithms.

g) Disease prognosis: Models that forecast how different medical problems will evolve may be created using machine learning methods. Machine learning algorithms, for instance, can forecast the prognosis of a certain malignancy or the chance of a disease recurrence.

h) Clinical decision assistance: Healthcare professionals can receive clinical decision help from machine learning algorithms. Machine learning models

can assist healthcare practitioners in making better decisions regarding patient care by evaluating patient data and making suggestions based on that data.
i) Remote patient monitoring: Wearable technology and smartphone apps, for example, may be utilised to remotely monitor patients using machine learning algorithms. Machine learning algorithms can notify healthcare practitioners of any issues before they get serious by evaluating patient data in real-time.
j) Patient stratification: Using their medical history or risk factors, patients can be grouped according to machine learning algorithms. This can assist medical professionals in identifying individuals who are more likely to experience negative outcomes and in allocating their care accordingly.

To safeguard patient confidentiality while using sensitive health data, strong privacy and security measures are needed. Some machine learning models, especially those involving deep learning techniques, can be difficult to understand and comprehend. In healthcare, efforts are still being made to improve the interpretability of ML models [6]. To ensure dependability and efficacy in actual healthcare settings, extensive validation is necessary before integrating machine learning predictions into clinical practice.ML algorithms are used in the creation of risk score models, which determine a person's probability of acquiring diabetes based on a number of variables. To improve diabetes care, real-time feedback and glucose level prediction are provided by continuous glucose monitoring devices using machine learning [7]. Also, machine learning algorithms examine genomic data to find genetic markers linked to a person's vulnerability to diabetes, which helps to improve prediction accuracy.

The ability of machine learning to spot patterns and links that human analysts may not instantly see is one of the technology's key benefits in the study of medical data. The ability of traditional statistical approaches to identify novel and unexpected associations in the data is constrained by their reliance on a priori assumptions and preconceptions. Machine learning algorithms, on the other hand, are able to analyse enormous volumes of data, spot patterns and trends, and produce precise forecasts. In order to identify early indicators of cancer that a human observer would overlook, machine learning algorithms, for instance, can evaluate medical imaging data. In a similar vein, machine learning may examine genetic data to find novel biomarkers for the detection and management of disease.

By forecasting patient outcomes based on a variety of clinical and biological parameters, machine learning algorithms may also be utilised to improve medical diagnosis and therapy. Machine learning algorithms can create individualised treatment regimens that maximise treatment efficacy and lower the likelihood of side events by evaluating patient data and seeing trends in the course of the disease. For instance, by examining electronic medical data, machine learning algorithms can identify patients who are more likely to experience difficulties and take action before they do. In a similar vein, machine learning algorithms are able to use genetic data analysis to forecast an individual's likelihood of contracting specific diseases and create tailored preventative plans.

Large volumes of data from clinical trials, post-market surveillance, and preclinical research may be analysed using machine learning algorithms to find possible therapeutic targets and streamline the drug development process. Machine learning systems, for instance, are able to predict therapeutic efficacy and find possible pharmacological targets by analysing molecular data [8]. Similar to this, machine learning algorithms may scan data from clinical trials and electronic health records to pinpoint patient groups most likely to benefit from a given medication and to customise dosage and treatment plans.

Even if machine learning has a lot of potential for analysing medical data, there are a number of obstacles and restrictions that need to be overcome to guarantee the ethical and safe application of these algorithms in clinical settings. The requirement for transparent and understandable algorithms is one of the primary obstacles. Machine learning algorithms are typically opaque, making it difficult to comprehend how they make their predictions [9]. Healthcare professionals find it difficult to trust the algorithms and apply them to clinical decision-making because of this lack of openness. Building confidence in machine learning algorithms and guaranteeing their safe and moral application in medical practice requires interpretable algorithms that can shed light on how they make their predictions [10].

1.6 CHALLENGES IN DIABETES EARLY DIAGNOSIS USING CONVENTIONAL METHODS

Although traditional methods have been vital in identifying those who are at risk of diabetes, they have drawbacks that may reduce the efficacy of early diagnosis [18] [19]. For the purpose of creating plans to enhance early detection techniques, it is imperative to comprehend these obstacles. The following are the main problems with traditional methods (Table 1.1):

TABLE 1.1
Challenges in Diabetes Early Diagnosis Using Traditional Methods

S.No	Challenges	Issues	Impact
1	Limited Sensitivity and Specificity	Conventional methods may lack the sensitivity to detect subtle changes in early stages of diabetes, leading to false negatives. Similarly, they might lack specificity, resulting in false positives.	Inaccurate results can lead to delayed intervention for those at risk or unnecessary concern and further testing for individuals incorrectly identified as high risk.
2	Single Time Point Assessments	Many conventional approaches involve single time point assessments, such as fasting glucose tests or oral glucose tolerance tests. These assessments may not capture fluctuations or trends in glucose metabolism over time.	Dynamic changes in glucose regulation may be missed, delaying the identification of individuals at risk in the early stages of diabetes.

(*continued*)

TABLE 1.1 (Continued)
Challenges in Diabetes Early Diagnosis Using Traditional Methods

S.No	Challenges	Issues	Impact
3	Reliance on Laboratory Tests	Traditional methods heavily rely on laboratory tests, such as blood glucose measurements and HbA1c tests, which may not be easily accessible or feasible for regular monitoring.	Accessibility issues can limit the frequency of assessments, potentially missing crucial windows for early detection and intervention.
4	Limited Integration of Multiple Data Types	Conventional approaches often focus on a limited set of clinical and physiological parameters, neglecting the integration of diverse data types such as genetic, lifestyle, and behavioural information.	The holistic understanding of an individual's risk profile is compromised, potentially overlooking important contributors to diabetes risk.
5	Inability to Predict Progression to Diabetes	Conventional methods may not effectively predict the progression from prediabetes to diabetes, as they often focus on diagnosing established diabetes.	Early interventions to prevent the progression to diabetes may be delayed, missing opportunities for targeted preventive strategies.
6	Limited Emphasis on Behavioural Factors	Traditional approaches may not sufficiently emphasise lifestyle and behavioural factors that contribute to diabetes risk, such as diet, physical activity, and smoking habits.	Modifiable risk factors may go unaddressed, limiting the effectiveness of preventive measures focused on lifestyle modifications.
7	Challenges in Identifying Individuals with Genetic Predisposition	While family history is considered in conventional risk assessments, identifying individuals with specific genetic predispositions may be challenging without targeted genetic testing.	Some individuals at high genetic risk may not be identified through conventional methods, missing opportunities for early intervention.
8	Limited Patient Engagement and Adherence	Conventional approaches may face challenges in engaging patients and ensuring adherence to recommended testing protocols, particularly for asymptomatic individuals.	Reduced patient engagement can lead to missed opportunities for early detection and intervention in high-risk individuals.

1.7 MACHINE LEARNING'S POTENTIAL TO IMPROVE DIABETES PREDICTIVE ACCURACY AND EFFICIENCY

Because machine learning (ML) provides sophisticated analytical tools that may greatly improve predicted accuracy and operational efficiency, it has enormous potential to change the diabetes detection environment [11]. Here is a look at the benefits machine learning (ML) offers the field (Table 1.2):

TABLE 1.2
Machine Learning's Potential to Improve Diabetes Predictive Accuracy and Efficiency

S.No	Potential	Issues	Impact
1	Use of Diverse Data Sources	Machine learning algorithms have the ability to easily integrate and analyse a variety of data sources, such as genetic data, electronic health records, lifestyle data, and real-time patient monitoring.	ML models may find intricate patterns and correlations that traditional methods would miss by taking into account a wide variety of data, which results in more precise predictions.
2	Early Subtle Pattern Detection	ML has the potential to identify early markers of diabetes even prior to the onset of clinical symptoms since it is highly skilled at identifying complex and subtle patterns in huge datasets.	Timely intervention made possible by early identification may be able to stop the progression of prediabetes to diabetes and lower the risk of complications.
3	Personalised Risk Stratification	By taking into account individual variances in genetics, lifestyle, and medical history, machine learning algorithms can provide personalised risk profiles.	By tailoring interventions to each person's specific risk factors, this personalised approach maximises the efficacy of preventative measures.
4	Constant Learning and Adaptability	Assures that forecasts change in response to fresh data and study since machine learning models are able to continuously learn from and adapt to new knowledge.	Predictive models become more relevant and long-lasting as a result of their flexibility, which also makes them more resistant to shifts in patient demographics and healthcare environments.

(continued)

TABLE 1.2 (Continued)
Machine Learning's Potential to Improve Diabetes Predictive Accuracy and Efficiency

S.No	Potential	Issues	Impact
5	Enhanced Dimensionality Reduction and Feature Selection	ML algorithms choose the most pertinent variables for prediction by using advanced feature selection and dimensionality reduction techniques.	Machine learning (ML) models increase interpretability of prediction models, decrease computing complexity, and boost efficiency by concentrating on important features.
6	Including Time-Series Data Integration	ML can analyse trends and changes in health metrics over time by handling time-series data well.	This skill is critical for recording dynamic changes in health status, particularly for diseases like diabetes where effective prediction depends on longitudinal data.
7	Predictive Analytics for Progression Risk	Actionable insights for early treatments can be obtained by using machine learning (ML) models to evaluate the risk of diabetes progression from prediabetes [20].	Targeted preventive approaches, such as pharmaceutical therapies and lifestyle interventions, can be used to reduce the chance of acquiring diabetes by predicting the progression risk.
8	Improved Model Interpretability	ML models in the healthcare industry will be easier to interpret, meaning that predictions will be clear and intelligible to medical experts.	By fostering trust in the prediction models, improved interpretability motivates healthcare professionals to use ML-based insights to clinical decision-making.
9	Real-time Decision help	By quickly analysing and interpreting data, machine learning models may offer healthcare practitioners real-time decision help [21].	By enabling prompt interventions and individualised patient care, real-time insights enable doctors to make well-informed decisions.
10	Affordable and Scalable Solutions	ML-based solutions provide the potential to be both affordable and scalable, which opens the door to their wider adoption in healthcare environments [22].	By enabling the integration of ML-enhanced predictive models into standard clinical procedures, scalability guarantees that a broader population can be reached and healthcare resources are optimised.

(Continued)

1.8 SOURCES OF DATA FOR DIABETES PREDICTION

Predictive modelling for diabetes involves the analysis of diverse datasets to identify risk factors, patterns, and early indicators of the condition. Leveraging various sources of data provides a comprehensive understanding of an individual's health status. Here are key sources of data for diabetes prediction (Table 1.3).

TABLE 1.3
Sources of Data for Diabetes Prediction

S.No	Source	Data Type	Significance
1	Electronic Health Records (EHRs):	Clinical and medical history, laboratory results, medication records.	Significance: EHRs provide a longitudinal view of a patient's health, enabling the tracking of trends, comorbidities, and changes in health parameters over time.
2	Genetic and Genomic Data	Genetic markers, family history, genomic variations.	Genetic information contributes to understanding an individual's predisposition to diabetes and can inform risk assessments. Advances in genomic medicine enhance predictive accuracy.
3	Demographic Data	Age, gender, ethnicity, socioeconomic status.	Demographic information helps in risk stratification and understanding population-specific patterns in diabetes prevalence and risk factors.
4	Lifestyle and Behavioural Data:	Data Type: Diet, physical activity, smoking habits, alcohol consumption.	Lifestyle factors play a crucial role in diabetes risk. Data on behaviour and lifestyle help tailor interventions and preventive strategies.
5	Anthropometric Measurements:	Body mass index (BMI), waist circumference, waist-to-hip ratio.	Anthropometric data provides insights into obesity and central adiposity, which are risk factors for Type 2 diabetes.
6	Physiological and Clinical Measurements	Blood pressure, cholesterol levels, triglycerides, HbA1c.	Physiological and clinical measurements offer indicators of metabolic health and contribute to the assessment of diabetes risk and management.
7	Continuous Glucose Monitoring (CGM) Data	Real-time glucose levels, glucose variability.	CGM data provides a dynamic and continuous assessment of glucose levels, offering valuable information for diabetes prediction and management.
8	Mobile Health (mHealth) and Wearable Device Data	Physical activity tracking, sleep patterns, heart rate.	Wearable devices and mHealth apps collect real-time data, contributing to a more holistic understanding of an individual's lifestyle and health behaviours.

(continued)

TABLE 1.3 (Continued)
Sources of Data for Diabetes Prediction

S.No	Source	Data Type	Significance
9	Environmental and Geographical Data	Geographic location, environmental factors [27].	Significance: Environmental data, including air quality and access to healthcare resources, may contribute to understanding regional variations in diabetes prevalence.
10	Social Determinants of Health Data	Education level, employment status, housing conditions.	Social determinants of health influence diabetes risk. Incorporating this data helps address barriers to care and promotes health equity.
11	Dietary Patterns and Nutritional Data	Dietary intake records, nutritional assessments.	Understanding dietary patterns and nutritional intake contributes to assessing diabetes risk and developing personalised dietary recommendations.
12	Medication History	Prescription history, medication adherence.	Medication history provides insights into the management of diabetes and other comorbid conditions, influencing predictive models.
13	Patient-Reported Outcomes (PROs)	Patient-reported symptoms, quality of life assessments.	PROs offer subjective insights into an individual's well-being, contributing to a more comprehensive health assessment.
14	Machine Learning-Generated Features	Predictive features generated by machine learning models.	Machine learning algorithms may identify novel features and patterns that contribute to predictive accuracy.
15	Population Health Data	Aggregated health data at the population level.	Population health data contribute to understanding broader trends, risk factors, and preventive strategies in diabetes.

1.9 PREPROCESSING STEPS TO HANDLE MISSING DATA AND ENSURE DATA QUALITY IN DIABETES PREDICTION

Managing incomplete data and guaranteeing data integrity are essential phases in developing a trustworthy diabetes prediction model. To resolve missing data and improve the dataset's overall quality, a number of preprocessing techniques can be used.

1. Finding Missing Data: Start by figuring out which values in the dataset are missing. To find null values and comprehend how they are distributed among features, use Python tools such as pandas.
2. Imputation Methods: Select the most suitable imputation methods to replace any missing data. For numerical values, common imputation techniques

include mean, median, or mode; for categorical features, the most frequent category is used. More sophisticated techniques like regression imputation or k-nearest neighbours (KNN) may also be taken into account [28].
3. Data Transformation: To guarantee consistent scaling across characteristics, use normalisation or standardisation. This lessens the chance that differences in the size of certain features may introduce bias into the model.
4. Outlier Detection and Handling: Identify and handle outliers that may negatively influence the model. To identify and deal with outliers effectively, apply machine learning algorithms or statistical techniques.
5. Feature Engineering: To enhance the performance of the model, add new, significant features or modify current ones. Predictive power can be increased, for instance, by grouping ages or by developing interaction terms between pertinent characteristics [29].
6. Execute comprehensive quality control procedures to find and fix mistakes or discrepancies in the dataset. This entails making sure that all submissions follow the required data types and verifying data integrity and duplicate records.
7. Data Splitting: To evaluate the model's performance on untested data, divide the dataset into training and testing sets. This contributes to confirming the model's generalizability [30].

1.10 MODEL BUILDING FOR SELECTING CORRECT ML ALGORITHM

1.10.1 Model Selection

Taking into account a variety of classification techniques, early diabetes prediction is basically a classification problem. Support vector machines, decision trees, random forests, logistic regression, and gradient boosting are typical options. These algorithms work well for situations involving binary classification, where the result is either non-diabetic or diabetic.

1. A linear model used for binary classification is called logistic regression. Through the application of a logistic function, it forecasts the likelihood that an instance will belong to a specific class [30].
2. Decision Trees: To create a structure like a tree, Decision Trees iteratively divide data according to characteristic. A class prediction is represented by each leaf node.
3. Random Forests: Random Forests are an ensemble of decision trees. To increase accuracy and reduce overfitting, they construct many trees and merge their predictions.
4. Support Vector Machines (SVM): SVM uses a high-dimensional space to identify the best hyperplane for optimally separating classes in order to classify data.

5. Gradient Boosting: Gradient Boosting creates an ensemble of weak learners in a stepwise manner, with each member fixing the mistakes of the preceding one to create an extremely potent prediction model.
6. Early diabetes diagnosis prediction is a crucial endeavour with important public health consequences. A number of factors, including ensemble techniques, assessment metrics, cross-validation, hyperparameter tweaking, and interpretability, must be carefully taken into account while creating a trustworthy prediction model [31].

1.10.2 Cross-Validation

As data for early diabetes prediction is frequently scarce and valuable, cross-validation becomes essential to the model-building process. Partitioning the dataset into k subsets, training the model on k-1 subsets, and verifying it on the remaining subset are the conventional approaches, such as k-fold cross-validation. After k repetitions of this process, the average performance is evaluated. This method is especially crucial for reducing the chance of overfitting, accurately estimating the model's capacity for generalisation, and reliably assessing the model's performance across various data subsets [31]. Cross-validation is useful in determining if the model can effectively adjust to different patterns and subtleties in the data, given the complexity and multivariate nature of diabetes. It is important to make sure that the model learns fundamental patterns that generalise effectively to fresh, unknown data rather than just memorising the training set in order to forecast diabetes early on. For a model to be useful in the actual world, it must be able to make accurate predictions. Cross-validation offers a thorough evaluation of this stability.

1.10.3 Hyperparameter Tuning

A crucial stage in the creation of machine learning models for early diabetes prediction is hyperparameter tweaking. Hyperparameters are the model configuration settings that are predetermined before training starts and are not learnt from the data. Hyperparameters in the context of diabetes prediction might include things like the support vector machine's kernel selection, regularisation strength, or learning rate. A methodical investigation of various values is required to determine the combination that optimises the model's performance while tuning these hyperparameters. For instance, determining the ideal learning rate guarantees that the model converges well during training, and using the right regularisation can aid in preventing overfitting, particularly in situations with sparse data. Hyperparameter tweaking is essential for calibrating the model to attain optimal performance in early diabetes prediction, where the interactions between features and diabetes risk may be complex [32]. In order to assess various combinations of hyperparameter values, grid search and randomised search techniques are frequently used in the process of hyperparameter tuning. Iterative processes like these need balance and careful thought since too complicated models with lots of hyperparameters might cause overfitting, while too basic models might not be able to capture the nuances of early diabetes signs.

Machine Learning for Diabetes: From Data Prep to Early Prediction 19

1.10.4 Ensemble Methods

An effective way to improve the early diabetes models' predicting ability is to use ensemble methods. Ensemble techniques like Gradient Boosting and Random Forests can be quite helpful in this situation. The way these techniques work is by building several base models and merging their predictions to increase overall resilience and accuracy. For example, Random Forests generate several decision trees during training and then aggregate their forecasts. This method works well for identifying a variety of patterns in the data, lowering the chance of overfitting, and enhancing the generalisation abilities of the model. Ensemble approaches are excellent in capturing the complexities of the correlations that may exist between different health indicators and the risk of diabetes in the context of early diabetes prediction, where these relationships may be nonlinear and complicated [33]. Another ensemble technique is gradient boosting, which constructs a sequence of weak learners one after the other with the goal of fixing the faults of the combined ensemble. The model may adjust to the complexities of the data and increase in predicted accuracy over time thanks to this iterative process. In situations where individual models would find it difficult to capture the entire range of early diabetes indications, ensemble approaches might be very helpful.

1.10.5 Metrics for Evaluation

Accurately evaluating the efficacy of early diabetes prediction algorithms depends on selecting the right assessment measures. The precise objectives of the prediction job and the relative significance of various forms of prediction mistakes in the context of medicine determine which metrics are used.

Metrics that are often employed include:

a) Accuracy: The percentage of cases out of all instances that were successfully predicted. Although accuracy makes sense, it could not be enough in datasets that are unbalanced, meaning that one type occurs far more frequently than the other.
b) Precision can be defined as the ratio of accurately predicted positive observations to the total number of positive predictions. When reducing false positives is crucial, like in healthcare applications, precision is useful [32].
c) Recall (Sensitivity): The ratio of accurately anticipated positive observations to the total actual positives. When the objective is to minimise false negatives while capturing the greatest number of real positives, recall plays a crucial role.
d) The harmonic mean of recall and accuracy is the F1-Score. Because it strikes a balance between recall and accuracy, the F1-Score is especially helpful in situations when there is a disparity between the classes.
e) The model's ability to distinguish between positive and negative occurrences across a range of threshold values is evaluated using the area under the receiver operating characteristic curve, or AUC-ROC. Understanding the model's total discriminatory power may be gained by utilising AUC-ROC.

The assessment criteria selected in the context of early diabetes prediction are determined by the particular goals of the prediction job as well as the possible repercussions of false positives and false negatives. False negatives, for example, may result in missed chances for early intervention, while false positives in the healthcare industry may lead to needless treatments.

1.10.6 Interpretability

Early diabetes prediction models must be interpretable, especially in the medical field where choices can directly affect the health of patients. Models that allow for easy comprehension and explanation of the correlations between input features and predictions are known as interpretable models. Interpretability guarantees that medical practitioners can rely on and understand the model's decision-making process in the context of early diabetes prediction. Simpler models, such as decision trees or logistic regression, are by nature easier to understand than more intricate models, like neural networks. Clear insights into the decision logic of the model are provided by the coefficients that are produced by logistic regression, which show the influence of each parameter on the estimated chance of diabetes. Conversely, decision trees are only sets of if-else criteria that are simple to understand. For predictive models to be accepted and used in clinical practice, interpretability is essential. It is imperative that healthcare practitioners comprehend not just the accuracy of the model's predictions, but also the reasoning behind them. Building confidence in the model's skills and making sure it is in line with current medical knowledge depend on its openness [3]. An important factor is to strike a balance between interpretability and prediction performance. Even though complicated models may produce results with more accuracy, it can be difficult to understand how they make decisions. Finding the ideal mix requires choosing a model that offers clear insights into the variables impacting early diabetes forecasts while also being accurate.

1.11 CONCLUSION

In this book chapter, we have explored the possibilities of machine learning algorithms for diabetes prediction for early intervention and better patient outcomes. The development of an effective early diabetes prediction model involves a multi-faceted approach. Cross-validation ensures the model's robustness and generalizability. Hyperparameter tuning optimises the model's configuration for the specific characteristics of the diabetes dataset. Ensemble methods enhance predictive accuracy by leveraging diverse learning patterns. Careful selection of evaluation metrics guides model assessment based on the specific goals of the prediction task. Interpretability is crucial for gaining insights into the model's decision-making process, making it more usable and trustworthy in a clinical context. By systematically addressing these elements, the resulting predictive model can offer valuable insights and aid in early intervention for individuals at risk of developing diabetes. Through a thorough understanding of data preparation, model creation, and deployment techniques, the chapter is an invaluable tool for academics, practitioners, and healthcare professionals who want to use cutting-edge technology to improve diabetes treatment.

REFERENCES

[1] Alex, S.A., Nayahi, J., Shine, H., Gopirekha, V. Deep convolutional neural network for diabetes mellitus prediction. Neural Comput. Appl. 2022;34:1319–1327. https://doi.org/10.1007/s00521-021-06621-6

[2] Zou, Q., Qu, K., Luo, Y., Yin, D., Ju, Y., Tang, H. Predicting diabetes mellitus with machine learning techniques. Front. Genet. 2018;9:515. https://doi.org/10.3389/fgene.2018.00515

[3] Afsaneh, E., Sharifdini, A., Ghazzaghi, H., Ghobadi, M.Z. Recent applications of machine learning and deep learning models in the prediction, diagnosis, and management of diabetes: A comprehensive review. Diabetol. Metab. Syndr. 2022;14:196. https://doi.org/10.1186/s13098-022-00919-8.

[4] Nagarajan, S., Chandrasekaran, R.M. Design and implementation of expert clinical system for diagnosing diabetes using data mining techniques. Indian J. Sci. Technol. 2015;8(8):771–776. https://doi.org/10.17485/ijst/2015/v8i8/69272

[5] Oladimeji, O. Machine learning in smart health research: A bibliometric analysis. International Journal of Information Science and Management. 2023;21:119–128. https://doi.org/10.22034/ijism.2022.1977616.0

[6] Sachdeva, S., Ali, A., Khalid, S. Telemedicine in healthcare system: A discussion regarding several practices. In: Choudhury, T., Katal, A., Um, J.S., Rana, A., Al-Akaidi, M. (eds.) Telemedicine: The Computer Transformation of Healthcare. TELe-Health. Cham: Springer; 2022. https://doi.org/10.1007/978-3-030-99457-0_19

[7] Jujjavarapu, C., Suri, P., Pejaver, V. et al. Predicting decompression surgery by applying multimodal deep learning to patients' structured and unstructured health data. BMC Med. Inform. Decis. Mak. 2023;23(2). https://doi.org/10.1186/s12911-022-02096-x

[8] Mohsen, F., Ali, H., El Hajj, N. et al. Artificial intelligence-based methods for fusion of electronic health records and imaging data. Sci. Rep. 2022;12:17981. https://doi.org/10.1038/s41598-022-22514-4

[9] Hobensack, M., Song, J., Scharp, D., Bowles, K., Topaz, M. Machine learning applied to electronic health record data in home healthcare: A scoping review. *Int. J. Med. Inf.* 2022;170:104978. https://doi.org/10.1016/j.ijmedinf.2022.104978

[10] Sachdeva, S., Ali, A., Khan, S. Secure and privacy issues in telemedicine: Issues, solutions, and standards. In: Choudhury, T., Katal, A., Um, JS., Rana, A., Al-Akaidi, M. (eds) Telemedicine: *The Computer Transformation of Healthcare*. TELe-Health. Springer; 2022. https://doi.org/10.1007/978-3-030-99457-0_21

[11] Mustafa, A., Rahimi Azghadi, M. Automated machine learning for healthcare and clinical notes analysis. *Computers* 2021;10:24. https://doi.org/10.3390/computers10020024

[12] www.prevention.com/health/health-conditions/a21764231/type-2-diabetes-definition/

[13] Mantas, J. Setting up an easy-to-use machine learning pipeline for medical decision support: A case study for COVID-19 diagnosis based on deep learning with CT Scans. Importance Health Inform. Public Health Pandemic. 2020;272:13.

[14] Yala, A., Lehman, C., Schuster, T., Portnoi, T., Barzilay, R. A deep learning mammography-based model for improved breast cancer risk prediction. Radiology. 2019;292:60–66. https://doi.org/10.1148/radiol.2019191194

[15] Malasinghe, L.P., et al. Remote patient monitoring: A comprehensive study. J. Ambient Intell. Human Comput. 2019;10:57–76. https://doi.org/10.1007/s12652-017-0598-x

[16] Tamarai, K., Bhatti, J.S., Hemachandra Reddy, P. Molecular and cellular bases of diabetes: Focus on type 2 diabetes mouse model-TallyHo. *Biochim. Biophys. Acta* 2019;1865(9):2276–2284, ISSN 0925-4439. https://doi.org/10.1016/j.bbadis.2019.05.004www.sciencedirect.com/science/article/pii/S0925443919301644

[17] Diabetes Types and Prevalence in Children in India. Civils Daily. Accessed from www.civilsdaily.com/news/diabetes-types-children-india/

[18] Huang, SC., Pareek, A., Seyyedi, S. et al. Fusion of medical imaging and electronic health records using deep learning: A systematic review and implementation guidelines. *NPJ Digit. Med.* 2020;3:136. https://doi.org/10.1038/s41746-020-00341-z

[19] Kaur, I. and Ali, A. (2024). A Complete Study on Machine Learning Algorithms for Medical Data Analysis. In Fog Computing for Intelligent Cloud IoT Systems (eds C. Banerjee, A. Ghosh, R. Chakraborty and A.A. Elngar). https://doi.org/10.1002/9781394175345.ch7

[20] I. Kaur and A. Ali, "An In-Depth Exploration of Machine Learning Algorithms and Performance Evaluation Approaches for Personalized Diabetes Prediction," 2024 International Conference on Emerging Innovations and Advanced Computing (INNOCOMP), Sonipat, India, 2024, pp. 532-538, doi: 10.1109/INNOCOMP63224.2024.00093.

[21] A. Khanna, P. Singh, Inzimam-Ul-Hassan and I. Kaur, "Predictive Analytics for Cardiovascular Health: A Machine Learning Approach," 2024 International Conference on Advances in Modern Age Technologies for Health and Engineering Science (AMATHE), Shivamogga, India, 2024, pp. 1-6, doi: 10.1109/AMATHE61652.2024.10582101.

[22] Kushwaha, S., Bahl, S., Bagha, A.K., Parmar, K.S., Javaid, M., Haleem, A., Singh, R.P. Significant applications of machine learning for COVID-19 pandemic. J. Ind. Integr. Manag. 2020;5(4):1–10. https://doi.org/10.1142/S2424862220500146.

[23] Lalmuanawma, S., Hussain, J., Chhakchhuak, L. Applications of machine learning and artificial intelligence for COVID-19 (SARS-CoV-2) pandemic: A review. Chaos Sol. Fract. 2020;110059. https://doi.org/10.1016/j.chaos.2020.110059.

[24] M. Sharma, I. Kaur, G. Saini, V. Thakur and A. Mishra, "Predictive Modeling for Stroke Risk Assessment Using Machine Learning," 2024 International Conference on Intelligent Systems for Cybersecurity (ISCS), Gurugram, India, 2024, pp. 1-6, doi: 10.1109/ISCS61804.2024.10581155.

[25] Mehta, V., Bawa, S., Singh, J. Analytical review of clustering techniques and proximity measures. *Artificial Intelligence Review*. 2020;53(1S):29. https://doi.org/10.1007/s10462-020-09840-7

[26] Nilashi, M., Ibrahim, O.B., Ahmadi, H., Shahmoradi, L. An analytical method for diseases prediction using machine learning techniques. Comput. Chem. Eng. 2017;106:212–223. https://doi.org/10.1016/j.compchemeng.2017.02.021.

[27] Perveen, S., Shahbaz, M., Keshavjee, K., Guergachi, A. Metabolic syndrome and development of diabetes mellitus: Predictive modeling based on machine learning techniques. IEEE Access 2018;7:1365–75. https://doi.org/10.1109/ACCESS.2018.2792741.

[28] Gökhan, S., Nevin, Y. Data analysis in health and big data: A machine learning medical diagnosis model based on patients' complaints. *Commun. Stat. Theory Methods.* 2019;1S:10.

[29] Zheng, T., Xie, W., Xu, L., He, X., Zhang, Y., You, M., Yang, G., Chen, Y. A machine learning-based framework to identify type 2 diabetes through electronic health records. Int. J. Med. Inform. 2017;97:120–127. https://doi.org/10.1016/j.ijmedinf.2016.10.001.

[30] Irfan, H., Raja, R., Anand, M., Karnatak, V., Ali, A. Comprehensive robustness evaluation of an automatic writer identification system using convolutional neural networks. *J. Autonomous Intell.* 2024;7(1):1–14. https://doi.org/10.32629/jai.v7i1.763.

[31] Yousef, R., Khan, S., Gupta, G., Albahlal, B.M., Alajlan, S.A., Ali, A. Bridged-U-net-ASPP-EVO and deep learning optimization for brain tumor segmentation. *Diagnostics* 2023;13(16):2633. https://doi.org/10.3390/diagnostics13162633

[32] Woldaregay A.Z., Årsand E., Botsis T., Albers D., Mamykina L., Hartvigsen G. Data-driven blood glucose pattern classification and anomalies detection: Machine-learning applications in type 1 diabetes. *J. Med. Internet Res.* 2019;21(5):e11030. https://doi.org/10.2196/11030

[33] Rao, S.R., Desroches, C.M., Donelan, K., Campbell, E.G., Miralles, P.D., Jha, A.K. Electronic health records in small physician practices: Availability, use, and perceived benefits. *J. Am. Med. Inform. Assoc.* 2011;18(3):271–275. https://doi.org/10.1136/amiajnl-2010-000010

2 Digital Image Forgery Techniques for Smart Information Generations

Gurmeet Kaur Saini and Salah Al-Majeed

2.1 INTRODUCTION TO DIGITAL IMAGE FORGERY

A new area involving machine learning and image processing is termed digital image forgery. The digital picture forgery is focused on image authentication and verification. By identifying forgeries, digital image forensics verifies the accuracy of the photographs [1]. The detection of photo manipulation is the goal of image forensics. The act of altering the original material with the intent to harm or make unauthorised changes to the original data is known as tampering [2]. This is made possible by the abundance of free software tools, which make it simple to create, edit, and change digital photos without revealing any illegal information [6]. Digital picture modifications involve resizing individual pixels or entire image blocks without displaying the altered image's impact. It is difficult for human eyes to see these changes [8]. Because of this, it is exceedingly challenging to determine whether the provided digital photographs are altered or tampered with [40].

The issue of digital picture forgery is becoming more prevalent in both criminal cases and public discourse on a daily basis. A growing area of study aimed at guaranteeing the validity and integrity of digital photographs is the detection of forgeries in them. Digital image alteration has been observed recently in courtrooms, fashion magazines, tabloid magazines, scientific journals, major media outlets, and picture frauds, among other places [3].

2.2 DIGITAL IMAGE FORGERY APPLICATIONS

- Cybercrime investigation is frequently employed in both private and criminal investigations.
- Forensic examination of the photos on social media websites.
- Utilised to identify altered or forged images.
- A method for detecting picture fraud is required in a number of areas, including copyright protection and preventing image manipulation or tampering. It is used in surveillance systems, multimedia security, digital forensic science, journalism, and other fields.

2.3 CLASSIFICATIONS OF APPROACHES

Techniques for detecting digital picture forgeries are separated into active and passive groups.

2.3.1 ACTIVE APPROACH

An image can be marked with details about the fraud, including a name, authorisation, etc. as part of an active detection method [22]. Furthermore, a distinct hardware implementation is required to verify the authenticity of the digital image.

2.3.1.1 Techniques of Active Approach:

a) Watermarking: This technique is employed to detect editing with digital images. The watermark needs to be included at the moment the image is created. A distinctive digital producer identifier (signature) on the content of photographs or videos is equivalent to adding a watermark on the digital image or video. After the image or video has been altered, this watermark will be removed, allowing the authorised recipient to examine it and confirm the accuracy of the data. Watermarking's objective is to conceal a message within a picture to ensure the image's copyright is protected and any message extracted from it is verified by comparing it to the original watermarks. The watermarks on the provided image won't change if it isn't altered. As a result, this approach depends on the information's original source. This method does not perform effectively with lossy compression since some images captured by cameras do not have watermarks in pictures as shown in Figure 2.1.

a.) Watermark symbol Image	b.) Actual Image
c.) Watermarked original Image	d.) Watermarked original image
(Watermark over the whole image)	(Watermark at the corner)

FIGURE 2.1 Example of Watermarking [33].

b) Digital Signatures: A digital signature is a mathematical procedure used in cryptography to confirm the validity of an original document [6]. Based on the contents, it creates a digital signature that includes the unique digital producer identification and

FIGURE 2.2 Signature Generator and Image Authentication Process [31].

pertinent content metadata. The signature is created by a producer and is unforgeable, much like a private key. As a result, the authenticator can confirm if the contents of an image or video match the data included in the signature to confirm the integrity of the received file as shown in Figure 2.2.

Concurrently generated are a digital image and a signature. This image's signature is an encrypted version that is kept apart from the rest of the file. Upon receiving an image, the recipient should first decrypt the signature and then compare the picture's hash codes to the appropriate values in the original signature. This image is considered "authentic" if they match [31].

2.3.1.2 Advantage of Active Approach:
- There is a lower computation cost.
- If one has knowledge of the original image, it is simple to grasp.

2.3.1.3 Disadvantage of Active Approach:
- These methods are not automatic; they require specific information about the original image, necessitating human intervention.
- Many digital photos available online lack digital signatures and watermarks, rendering active methods ineffective for verifying their authenticity [7].
- The transmission of signatures in a digital signature system requires additional bandwidth.

2.3.2 Passive Approach

This technique looks for picture forgeries without requiring embedded information in the images. The efficacy of the passive technique is contingent upon the remnants of the picture alteration process. The amount and location of picture forgery can also be ascertained using this method. There are two methods available to the passive technique to identify forgeries: picture source identification, which locates the equipment used to manipulate the digital image [12]. It indicates whether a computer or a digital camera was used to create the image. However, the location of the fake in the image cannot be determined by this method. Tampering detection: It identifies intentional picture alteration done with malicious intent. Here, altering an image with the intention of changing a portion of the original message is referred to as tampering [32].

2.3.2.1 Techniques of Passive Approach:

Passive techniques for detecting digital image forgeries operate without any pre-embedded information or additional hardware. They can be categorised as follows:

- **Pixel-based Techniques**: These methods detect inconsistencies at the pixel level.
- **Format-based Techniques**: These approaches identify statistical correlations resulting from specific compression techniques.
- **Camera-based Methods**: These methods analyse artifacts created by the sensor or camera's lens.
- **Physically-based Methods**: These techniques detect discrepancies in the three-dimensional interactions between real-world objects, light, and the camera using precise modeling and detection.
- **Geometry-based Techniques**: These methods assess the positioning and relations of real-world objects as captured by a digital camera [5].

2.3.2.2 Advantage of Passive Approach:

- Using an active method has no value for pre-existing photographs and data. This drawback was overcome by the passive technique, which allows pre-existing images to be provisioned as well [4].

2.3.2.3 Disadvantage of Passive Approach:

- The underlying premise of this method is that image alteration leaves no visible traces of forgery. As a result, this approach needs distinct image data. Thus, it is more sophisticated than an active strategy.

2.4 TYPES OF DIGITAL IMAGE FORGERY

Five main categories are used to define the forgeries.

a) Image Retouching
b) Image Splicing

c) Copy-Move (Cloning)
d) Morphing

a) Image Retouching: This type of image alteration is used to improve an image by removing unwanted elements and adding desired aspects. It also improves the image's quality to draw readers in. Using this technique, picture editors can alter the backdrop colour and the image colour by adding some eye-catching colours. The retouching example is displayed in Figure 2.3.

Using this technique, image editors alter the backdrop colour while simultaneously adding eye-catching colours to the original image to draw readers in.

b) Image Splicing: This technique involves copying specific portions of an original image and pasting them onto another image to merge different aspects from several images into a single image. The splicing border, which is displayed in Figure 2.4, can be found to identify this kind of forgery:

FIGURE 2.3 Image Retouching.

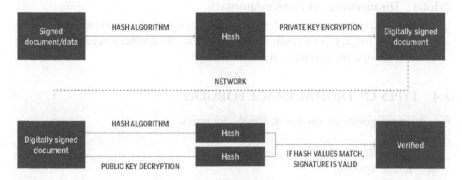

FIGURE 2.4 Image splicing.

Digital Image Forgery Techniques for Smart Information Generations 29

This illustration illustrates how parts from two separate images are blended to create the target image, in this case, the shark image and the helicopter from the base image.

c) Copy-Move: Using this method, you can copy and paste a section of an image to a different place within the same image. Usually, this tactic is used to hide some important information. The primary reason this counterfeit is used is that the replicated area's features are probably the same as the original, making it difficult for the human eye to detect. To decrease the impact of the original and pasted region, use the blur tool [23]. The copy-move example is displayed in Figure 2.5:

| Original picture | Edited Picture |

FIGURE 2.5 Copy-Move Image.

d) Morphing: This kind of media involves transforming an object from one image into another, giving each movie and image a distinct impact. This technique uses a smooth transition between two photos to replace the image of one person with that of another. The morphing example is displayed in Figure 2.6:

FIGURE 2.6 Image morphing.

This picture demonstrates how to seamlessly transition between two photos to transfer one person's image with another.

2.5 COPY-MOVE FORGERY

Copy-move forgery involves duplicating a part of an image and pasting it elsewhere within the same image. This type of tampering is often used to either conceal or add information. Since the duplicated areas can be very similar and hard to detect with the human eye, copy-move forgery detection techniques are designed to identify these subtle manipulations. These techniques often focus on finding similarities between copied and pasted elements in terms of noise, colour, and dynamic range [22, 39].

2.5.1 Copy-Move Forgery Detection Techniques

Modern methods for identifying picture forgeries that exclusively target copy-move forgeries are covered in this section.

Due to the nature of copy-move forgery—where copied and pasted portions are included in the same image—the majority of detection techniques take use of this characteristic. The presence of identical components is a sign of manipulation. Block-based methods and keypoint-based methods are the two categories into which these techniques can be divided [29].

2.5.1.1 Block-Based Methods

Block-based methods work by dividing the image into smaller, non-overlapping blocks and extracting feature vectors from each block [38]. The steps are as follows:

- **Image Partitioning**: The image is partitioned into fixed-size blocks.
- **Feature Extraction**: Feature vectors are extracted from each block.
- **Lexicographic Sorting**: These vectors are then sorted in a lexicographical order. Identical or similar vectors will appear consecutively in this sorted list.
- **Detection**: By comparing these sorted feature vectors, the method can identify duplicated regions. Any block pair with feature vectors below a certain difference threshold is flagged as potentially manipulated.

An advanced DCT (Discrete Cosine Transform)-based technique enhances this method. It involves:

- Calculating the DCT for each block.
- Applying a quantisation matrix to the DCT coefficients.
- Rounding the coefficients to the nearest integer and arranging them in a zigzag order to form a feature vector.
- Comparing the row feature vectors of the coefficients to detect tampering, where small differences indicate potential forgeries.

Digital Image Forgery Techniques for Smart Information Generations

2.5.1.2 Keypoint-Based Methods

Keypoint-based methods focus on identifying specific, distinctive points within the image (keypoints) [37] and use them to detect forgery. These methods are generally faster and more efficient because they handle fewer feature vectors and require less computation [36]. Two widely used keypoint-based techniques are:

- **Scale Invariant Feature Transform (SIFT)**: SIFT detects keypoints that are invariant to scale, rotation, and translation, making it robust against various transformations.
- **Speeded-Up Robust Features (SURF)**: SURF is an accelerated version of SIFT, which provides similar robustness but with faster processing times.

By employing these techniques, it becomes possible to effectively detect and identify copy-move forgeries within digital images.

2.6 OVERARCHING STRUCTURE FOR IMAGE FRAUD DETECTION SYSTEM

The goal of the Image Forgery Detection System (IFDS), a type of Pattern Recognition System (PRS), is to classify an input pattern (object) into one or more pre-established categories [43]. The pre-specified categories in IFDS are genuine and tampered with, and the item is an image [41].

The five processes of an IFD system include preprocessing, feature extraction, feature selection, classification, and assessment. Each component's output serves as an input for the following stage, which is depicted in Figure 2.7. Prior to selecting the most significant features, such as the matched ones, the input image must first be taken for preprocessing. From there, image features are retrieved. The classifier is then constructed using the features that were chosen. At last, the assessment is completed to determine the classifier's performance. Every component of the system needs to be properly planned in order to develop it with high detection accuracy and efficiency. These stages were briefly outlined in the sections that followed.

FIGURE 2.7 IFDS flowchart showing the key components [20].

2.6.1 Preprocessing

Preprocessing is the process of using picture enhancing methods to separate interesting patterns from the background or to lower noise in the data [13]. Furthermore, it involves converting the image across various colour systems. Actually, the process depends on the application.

2.6.2 Feature Extraction

Finding a new feature-based representation of the data (picture) is called feature extraction. Finding discriminant characteristics that accurately describe the data is the main concept. Good features must prevent redundancy and reduce the number of dimensions in the data.

- To draw out characteristics for image fraud detection, differencing approaches are employed [1], [2], and [14]. The subsequent subsections provide a review of a few of these methods.

Discrete Cosine Transform

The Discrete Cosine Transform (DCT) converts an image into a sum of cosine functions oscillating at different frequencies. This process changes the image representation from the spatial domain (intensity values) to the frequency domain (frequency coefficients). By highlighting the more visually significant information, the frequency domain offers a clearer understanding of the image's essential features. One of the key characteristics of DCT is its ability to compress data by concentrating most of the significant information into a few low-frequency coefficients [15–16]. This attribute makes DCT a popular choice for JPEG and other image compression formats. In practice, DCT transforms an array of pixel values into an array of frequency coefficients. The top-left coefficient, known as the DC component, captures the low-frequency data (overall average), while the remaining coefficients, known as AC components, represent the higher frequency details. The bottom-right coefficient denotes the highest frequency information. This transformation aids in efficient image compression by preserving essential details while reducing redundant data.

Dyadic Wavelet Transform (DyWT)

Discrete Wavelet Transform (DWT) offers both frequency and position information about the image, in contrast to DCT, which only provides frequency information. DWT is hence more favoured for usage in image processing. A multi-resolution representation of the image, including multidirectional features and an average (approximation), is produced by DWT. These representations are helpful in that they allow characteristics that are not detected at one resolution to potentially be detected at another [17].

DWT is a shift-invariant descriptor, which is a drawback despite its widespread use in copy-move fraud detection. It loses this characteristic as a result of the down sampling process used in the breakdown. Nevertheless, in copy-move fraud, the copied

Digital Image Forgery Techniques for Smart Information Generations 33

FIGURE 2.8 Change-invariant significance.

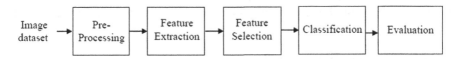

FIGURE 2.9 LBP code computation process.

and pasted portions might not line up with the two blocks' locations. Stated otherwise, these components might be moved. As a result, two distinct representations will be produced by the DWT descriptors that were taken from these blocks, missing the forging.

DyWT does not use down sampling, hence during the decomposition process, the image's size does not change. The duplicated and pasted stars are not in the same places in the two blocks. The two blocks' representations will be identical according to the shift-invariant descriptor and thus, the tampered part will be identified in Figure 2.8.

2.6.2.1 Local Binary Pattern (LBP)

LBP is a local operator that distinguishes between several texture kinds. Every pixel in a picture has a label (LBP code) defined by the original LBP operator [19]. A 3x3 neighbourhood of the pixel is thresholded by its intensity value in order to calculate the LBP code. The binary number '0' will be held by the neighbour if its pixel value is smaller than the centre; if not, it will hold '1'. To create a binary code, these neighbours' binary digits concatenate either clockwise or anticlockwise. The binary code's decimal value is known as the LBP code. The procedure of computing LBP codes is depicted in Figure 2.9.

2.6.3 Feature Selection

Feature selection is crucial in reducing the dimension and complexity of extracted features, achieved by eliminating redundant or unnecessary features. Techniques like correlation-based feature selection (CFS) and sequential forward selection (SFS) are commonly used for this purpose.

2.6.4 Classification

Classification involves assigning unknown data samples to predefined classes. This process includes training and testing phases. In the training phase, the system learns to map features extracted from a training set of images to predefined classes, creating a model or classifier.

During the testing phase, the system classifies the new images (i.e., testing set) using the learned model and the features that were extracted from them [42]. Depending on how many classes there are in the data, the classification can be either multiclass or two-class. Numerous techniques for classification exist; some of the more well-known ones are Support Vector Machines (SVM) [25], Neural Network (NN) [24], and Nearest Distance Classifier [23]. SVM technique is widely used in research for IFD [14]. Figure 2.10 illustrates the classification's training and testing stages:

a.) Support Vector Machine (SVM)

First, the Support Vector Machine (SVM) -A hyperplane that divides the data into two classes is defined by a Support Vector Machine (SVM) classifier. Vapnik and Corinna Cortes proposed SVM in 1995 [25]. There are two ways to determine the separating hyperplane: either the classes can be separated linearly or non-linearly (refer to Figure 2.11).

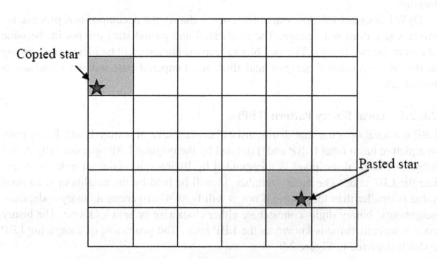

FIGURE 2.10 The training and testing phases of the classification.

Digital Image Forgery Techniques for Smart Information Generations 35

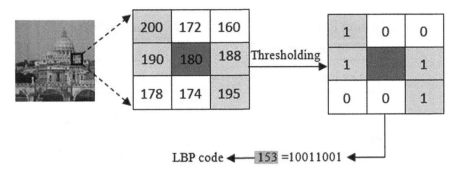

FIGURE 2.11 The two ways that the classes can be separated are (a) linearly and (b) non-linearly [30].

2.6.4.1 Linearly Separable Case

The samples of two distinct linearly separable classes are divided by an infinite number of hyper planes. The hyper plane with the maximum margin—that is, the maximum distance between the hyper plane and the closest samples—is the ideal one that improves the classifier's generalisation. The margin in Figure 2.12 is indicated by the distance between the dashed lines, and the samples that define this margin are referred to as support vectors. Finding the optimal hyperplane orientation to maximise the margin between support vectors is the general objective of support vector machines (SVMs).

2.6.4.2 Non-linearly Separable Case

The classes are not linearly separable in this instance. SVM uses kernel functions to handle this instead of curves to separate them. As shown in Figure 2.13, the kernel functions map the samples to a higher dimension space where the classes can be separated linearly. Although many kernel functions have been employed, the most often used ones are the polynomial, sigmoid, and Radial Basis Function (RBF) [26].

2.6.5 EVALUATION

The literature uses a variety of measurements to assess the classifier's performance. In the following subsections, a review of some of the most significant measurements used in IFD is provided.

Important terms that are required to comprehend the performance measurements include the following:

- The number of tampered images that are labelled as tampered is known as TP (True Positive).
- False Negative, or FN, is the quantity of altered photos that are accepted as real.

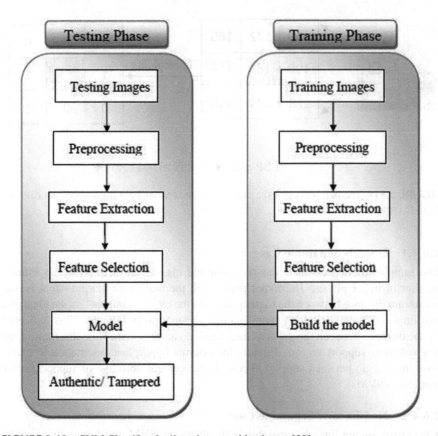

FIGURE 2.12 SVM Classifier for linearly separable classes [30].

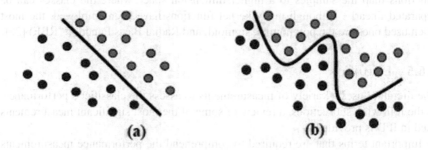

FIGURE 2.13 SVM for Non-linearly separable classes[30].

- The quantity of genuine photos that are deemed to be authentic is known as TN (True Negative).
- False Positive, or FP, is the quantity of real photos that are labelled as tampered ones.

Digital Image Forgery Techniques for Smart Information Generations 37

The parameters used to assess the performance are as follows [44]:

a.) Accuracy

The percentage of photos that the classifier correctly classifies is known as accuracy. The calculation is as [27]:

$$(TP + TN) / (TP + TN + FN + FP) \text{ equals accuracy.}$$

b.) True Positive Rate (TPR)

The percentage of real tampered images (positives) that are accurately identified as such is called the True Positive Rate (TPR), also referred to as Sensitivity or Recall. It comes out to [27]:

$$TP / (TP + FN) = TPR$$

c.) True Negative Rate (TNR)

The percentage of real, authentic images (negatives) that are accurately categorised as such is measured by True Negative Rate (TNR), also referred to as Specificity. The formula that follows is used to compute it [27]:

$$TN / (TN + FP) = TNR$$

d.) False Positive Rate (FPR)

The percentage of real, authentic images (negatives) that are incorrectly identified as tampered with is known as the False Positive Rate, or FPR. It is calculated as follows and equals 1 minus TNR:

$$FP / (FP + TN) = 1\text{-}TNR \text{ is TNR.}$$

e.) Error rate:

It calculates the proportion of photos that the classifier incorrectly classifies. It is computed as follows:

$$(FN + FP) / (TP + TN + FN + FP) \text{ is the error rate.}$$

2.7 LITERATURE SURVEY

Bayram, S. et al. [5] present a unique technique for detecting copy-move fraud in digital photos is presented by Bayram, S. et al. [5]. It is notably more resistant to lossy compression, scaling, and rotation sorts of manipulations. In order to reduce the computational complexity in identifying the duplicated picture regions, the authors also propose counting bloom filters as a substitute for lexicographic sorting, which is a common element of most suggested copy-move forgery detection techniques. The findings imply that the proposed features are quite good at detecting duplicated sections in the photos, even when the copied region has undergone significant image alteration. Moreover, it is seen that a little reduction in robustness is associated with a considerable increase in time efficiency when counting bloom filters are used.

Farid, H. et al. [14] evaluate the state of the art in this new and interesting field of visual picture since modern digital technology has begun to destroy the faith and confidence of integrity of digital image. Edited photos are becoming more common and sophisticated in a range of contexts, such as tabloid magazines, scholarly journals, courtrooms, fashion, mainstream media, political campaigns, and email hoaxes. Over the past five years, the science of digital forensics has evolved to restore some trust in digital photographs. In this study, the authors investigate several techniques to ensure image integrity.

Farid, H. et al. [13] offer a method to ascertain whether a piece of an image was originally compressed at a lesser quality than the rest. When making a digital fraud, it's often required to integrate various photos, like in the case of compositing one person's head onto another person's torso. If these photographs had different compression settings when they were first taken, there may be traces of the original JPEG compression quality in the digital composite. This method works with both low-quality and high-quality photos.

Image changes can be detected using a SIFT-based method as described by Ardizzone, E. et al. (2014). The three parts of the SIFT-based approach—texture analysis, cluster matching, and key point clustering—are compared after the SIFT-point matching technique is explained. The goal is to locate duplicates of the same object, or clusters of points, as opposed to searching for matching points. Cluster matching performs better in terms of results than single-point matching because it provides a thorough and coherent comparison between cloned objects. Lastly, textures of areas that match are compared and inspected in order to confirm results and eliminate false positives.

Bo X. et al. [8] introduced a method based on the SURF (Speed up Robust Features) descriptors, which are invariant to rotation, scaling, etc. due to the ease with which digital photos can be altered without creating obvious visual indications. Abuse of the modified images may result in unfavourable social, legal, or private outcomes. In order to do this, it is crucial and challenging to create effective methods for detecting digital photo forgeries, such as SURF. The research' findings demonstrate how well the recommended approach detects image region duplication and how resilient it is against additive noise and blurring.

Khan, S. et al. [24] discuss the blind image forensics technique for detecting copy-move forgeries. This method reduces the forged image's dimensions by using the DWT (Discrete Wavelet Transform). The compressed image is then divided into parts that overlap with a defined size. These blocks are sorted using lexicographic sorting, and duplicate blocks are found using a similarity criterion called phase correlation.

The duplication map, which is used to show discovered forgeries, provides the count of faked pixels.

Li, W., et al. [28] present a Fourier-Mellin transform with features extracted in the radius direction. To further reduce computational expenses, link processing is introduced to the counting bloom filters in place of hash value counting. Furthermore, a vector erosion filter is designed to cluster distance vectors together, a function that is commonly achieved by vector counters in pre-existing copy-move detection algorithms. Experimental data shows that the improved technique can effectively identify duplicated sections. Specifically, the improved approach can withstand repeated sections with a large rotation angle, whereas the present methods can only manage small rotations.

The challenge of identifying if an image has been altered is explored in Amerini, I., et al.'s work [2]. Specifically, the situation where a section of an image is copied and then pasted into another location to eliminate an unwanted aspect has been highlighted. Usually, the picture patch needs to be geometrically modified in order to fit the new environment. To identify these changes, a novel method based on scale invariant features transform (SIFT) is suggested. This method allows us to ascertain whether a copy-move attack has occurred and, if it has, to obtain the geometric transformation that was used to perform the cloning. Comprehensive experimental findings show that the methodology can reliably identify the modified region and estimate the geometric transformation parameters with high degree of confidence. It manages multiple cloning as well.

Bianchi, T., et al. [6] present a simple and reliable method for determining the presence of non-aligned double JPEG compression (NA-JPEG). When the DCT is computed using the grid from the previous JPEG compression, the approach depends on a single feature whose value depends on the integer periodicity of the DCT coefficients. Simple threshold detectors can classify NA-JPEG images more precisely than existing methods, especially in cases where the proposed feature is produced just with the use of DC coefficient statistics. For lesser picture sizes, this is also valid. Furthermore, this method can accurately determine the quantisation step and grid shift of the original JPEG compression, which can be applied to further investigate photographs that may have been altered.

Cao, G., et al. [9] propose a unique method for recognising unsharp masking (USM) sharpening procedures in digital photos. Overshoot artefacts can be seen in the sharpened photos close to side-planar edges. These relics can be a very useful feature for assessing the historical performance of the sharpening operation when they are picked up by a sharpening detector. Test findings on photographic images with respect to various sharpening operators show the value of this proposed approach.

Gharibi, F., et al. [16] describe a unique texture-based method to detect these kinds of forgeries in digital photographs. The proposed technique is to first partition the image into a few overlapping blocks and then use a modified Gabor filter to extract the feature vectors of each block. The collected feature vectors are then subjected to the PCA technique to reduce their dimension. Finally, in the matching phase, related or duplicate blocks are found using counting bloom filters. The experiment's results show that the proposed features are quite effective at correctly identifying the cloned portions, even when they have experienced lossy compression. As compared to other comparable works, the accuracy and performance of this method as well as a noticeable increase in detection rate illustrate the effectiveness of the proposed theory.

Guojuan, Z., et al. did a study on the digital watermarking method used in image forensics [17]. The authors began by assessing the current status of both national and international research, contrasting active and passive photo forensics, and using robust and fragile watermarking methods. The authors then listed three significant issues that still need to be fixed before digital watermarks may be used in image forensics. Finally, some suggestions are provided for further studies on forensic watermarks.

Huang, Y., et al. [21], proposed an enhanced DCT-based technique for identifying specific artifacts in images. Their method involves dividing the image into overlapping blocks of fixed sizes, each block then undergoes DCT to capture block-specific features. Dimensionality reduction is achieved through truncation of the feature vectors. These vectors are then sorted lexicographically, ensuring identical blocks are grouped together. This sorting facilitates the comparison of duplicate blocks during the matching process. The method introduces a strategy for determining the similarity between two feature vectors, enhancing its robustness. Experimental results validate the effectiveness of this approach in detecting repeated regions, even under conditions like additive white Gaussian noise, JPEG compression, or blurring.

The copy-move forgery detection approach discussed by Muhammad, N., et al. [30] is a non-intrusive and efficient method. The dyadic wavelet transform (DyWT) is used in this method to discover similarities and segment images. To determine how comparable the copied and pasted sections are structurally, DyWT and statistical measurements are employed. The results show that the suggested method outperforms the state-of-the-art procedures.

Redi, J. A., et al. [34] addressed two main issues: locating evidence of image forgeries and identifying the imaging instrument that took the picture. The field of digital picture forensics is growing in popularity among investigators due to the positive results of early research and its ever-growing range of applications. This survey is meant for academics and IT professionals who are interested in this area. It looks at current methods and provides an outlook on the history, present, and future of digital photo forensics.

Christlein, V., et al. [11] conducted a study to assess the effectiveness of various copy-move forgery detection techniques and processing steps, such as matching, filtering, outlier identification, and affine transformation estimations, under different post-processing scenarios. Their research aimed to evaluate the performance of existing feature sets by integrating them into a unified pipeline. They examined fifteen popular feature sets, analysing detection performance per picture and per pixel. To facilitate their study, the authors developed a comprehensive real-world copy-move dataset and a sophisticated software framework for systematic image alteration. Results indicated that keypoint-based Sift and Surf features, along with block-based DCT, DWT, KPCA, PCA, and Zernike features, exhibited exceptional performance.

These feature sets demonstrated high resilience against various noise sources and downsampling, reliably identifying duplicated regions.

Jing, L., et al. [23], explored the block matching approach and introduced a copy-move forgery detection technique based on local invariant feature matching. This method identifies copied and duplicated sections by comparing feature points, employing the Scale Invariant Transform method to extract local features and utilising k-d tree and Best-Bin-First methods for local feature matching. The proposed technique's computational complexity was found to be comparable to existing block matching algorithms, yet it achieved superior location precision. The method successfully detected sequentially copied and pasted portions, even after undergoing JPEG compression, Gaussian blurring, rotation, and scaling, as evidenced by testing outcomes.

Mahalakshmi, S. D., et al. published a technique for picture authentication that can identify digital image alterations [29]. This technique looks for typical image changes that are often done to modified photos, such as contrast enhancement, histogram equalisation, and resampling (rotation and scaling). Using the spectral signature approach to interpolation, one may determine rotation and rescaling and estimate parameters such as rotation angle and rescale factors. This rotation/rescaling detection technique perceives certain complete photographs as changed when JPEG images are compressed. By adding noise to the input photos, it also fixes this problem. The USC-SIPI database, which has generic, unedited photographs, was used to test this study, and the accuracy of the findings was determined to be good.

Qian, R., et al. [33] offer a new rotation-tolerant resampling detection method, on which an algorithm for blind photo fraud detection is built. To measure the difference in separation between two resampled pictures with different resampling histories, a "Rate-Distance" metric is created. Images are categorised using "Rate-Distances". Experimental findings show that the proposed technique can achieve high detection accuracy. In order to conceal any useful or significant information, it might be difficult to discern the changed region from the original image. For this reason, Siidevi, M., et al. [35] investigate a variety of image forgeries. The paper looks at how forged picture detection is currently done and assesses several copy-move detection techniques based on resilience and computational cost.

Amerini, I., et al. [3] presented a novel technique for identifying and localising copy-move forgeries. It is based on the J-Linkage technique that forms a robust clustering in the space of the geometric change. Experimental results on many datasets show that the proposed technique outperforms other comparable state-of-the-art strategies in terms of accuracy in the manipulated patch localisation and dependability in copy-move forgery detection.

Birajdar, G. K., et al. [7] provide an extensive bibliography on blind techniques for forgery detection and a summary of the most recent developments in the field of digital picture forgery detection. When employing passive or blind techniques, explicit previous knowledge of the picture is not required. Once they have been discovered, a generic structure for picture forgery detection methods is built. A summary of passive photo authentication is provided, along with a discussion of the existing techniques for blind forgery detection. The present state of the art in picture forgery detection techniques is discussed and a proposal for more study is offered.

Ferrara, P., et al. [15] give a comparison of two forensic techniques for the reverse engineering of a chain consisting of two JPEG compressions encircled by linear contrast enhancements. The second technique is based on the distribution of the first digit of DCT coefficients, whereas the first approach is based on the peak-to-valley behaviour of the double-quantised DCT coefficients' histogram. Parameter estimates for the processed chain under examination and chain detection have been added to these methods. More specifically, the proposed techniques provide an estimate of the quality factor and the linear contrast enhancement amount of the previous JPEG compression.

Hashmi, M. F., et al. [18] proposed a novel method using the Discrete Wavelet Transform (DWT) to detect photo tampering, specifically focusing on identifying picture copy-move forgeries. They utilise DWT for dimension reduction, dividing the image into four parts (LL, LH, HL, and HH) to apply the Scale Invariant Feature Transform (SIFT) technique, targeting the LL component with rich information. SIFT extracts significant features, generates descriptor vectors, and compares them to detect similarities, thereby discerning fake images and identifying visual copy-move forgeries. This approach not only identifies the forgery but also determines if image manipulation has occurred. On the other hand, Li, L., et al. [27] introduced an innovative method for copy-move forgery detection involving image filtering and division into overlapping circular chunks. They use rotation invariant uniform local binary patterns (LBP) to extract circular block features, enabling the identification of forged regions through matched blocks comparison. Experimental results validate the technique's robustness against JPEG compression, noise interference, blurring, flipping, and area rotation.

Panchal, P. M., et al. [31] provide two different methods for scale and rotation invariant interest point/feature detector and descriptor: Scale Invariant Feature Transform (SIFT) and Speed Up Robust Features (SURF). It also provides a way to recognise distinct invariant features in images, which may be used to consistently match different angles of an item or scene.

Cao, G., et al. [10] offer two novel methods for spotting contrast enhancement edits in digital images. First, as is customary in real-world applications, global contrast enhancement detection was applied to the JPEG-compressed pictures. The zero-height gap fingerprints are distinguished from the histogram peak/gap artefacts resulting from pixel value translations and JPEG compression using theoretical research. Finding the composite picture that was created by modifying the contrast of one or both source regions is the second step. The uniformity of regional artefacts is compared in order to identify composition borders and image frauds. Numerous experiments have verified the efficacy and efficiency of the recommended procedures.

Mohammad Farukh Hashmi [19] has researched on accelerated robust feature transform and wavelet transforms for copy-move picture fraud detection. A series of techniques combining accelerated robust feature transformations and wavelet transforms have been proposed by the research authors. The Speeded-Up Robust Feature (SURF), SURF in combination with the Discrete Wavelet Transform (DWT), and SURF in combination with the Dyadic Wavelet Transform (DyWT) have all been discussed by the writers. Unlike the previously proposed method, these approaches are used for the entire image in order to extract features, as opposed to the image being divided into blocks. The obtained results indicate that the recommended algorithms perform better than the alternatives in terms of computational complexity, rotation and scale invariance, and attack combination.

Jaberi, M., et al. [22] propose a method called MIFT that makes use of a more powerful collection of keypoint-based features that resemble SIFT features but are also invariant to mirror reflection alterations. The authors also recommend utilising an iterative method to gradually uncover more keypoint matches, which will improve the estimate of the affine transformation parameters and enhance the affine transformation. To reduce false positives and negatives during the extraction of the copied and pasted areas, the authors use morphological procedures, hysteresis thresholding, and "dense" MIFT features instead of typical pixel correlation. A comprehensive series of tests was conducted using a large dataset of real pictures in order to assess and contrast the proposed approach with rival options. Based on testing results, this technique is more accurate in identifying duplicated areas in copy-move photo forgeries, particularly in cases when the copied region is small.

Based on enhanced PCA-SIFT, Li, K., et al. [26] suggest an image forensics technique for identifying copy-move forgeries. The current method employs k-nearest neighbour to perform forgery detection after first extracting characteristics from an image and lowering its dimensionality. The similarities between the copied and pasted regions allow the descriptors to be compared, which is then used to look for any potential picture forgeries. According to extensive testing data, the system can accurately identify tampered images, evaluate their robustness and sensitivity to picture post-processing, and significantly increase time efficiency.

2.8 PROPOSED WORK

2.8.1 METHODOLOGY

In order to present a novel method for copy-move forgery detection, the image will first be converted using DWT into the wavelet domain, and then SIFT will be applied to the altered image in order to extract the features. SURF will be used for feature transformation at the second level. Features are increasingly prevalent when wavelet generates multispectral components. We will attempt to identify matching between these feature descriptors after getting the interest point feature descriptor in order to determine whether or not post-processing manipulation of the provided image has occurred. Our research demonstrates that the combination of SURF and SIFT features, with their strong performance and excellent computational efficiency, is the best option.

2.8.2 FLOWCHART

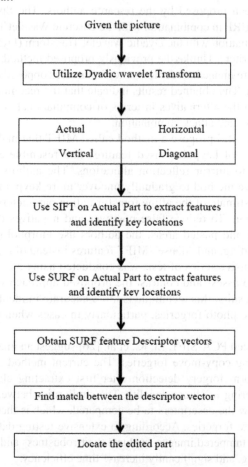

FIGURE 2.14 Diagram showing the planned system's flow.

REFERENCES

1. Alam, Sanawer, and Deepti Ojha. "A literature study on image forgery." *International Journal of Advance Research in Computer Science and Management Studies* 2, no. 10 (2014): 182–190.
2. Amerini, Irene, Lamberto Ballan, Roberto Caldelli, Alberto Del Bimbo, and Giuseppe Serra. "A sift-based forensic method for copy–move attack detection and transformation recovery." *IEEE Transactions on Information Forensics and Security* 6, no. 3 (2011): 1099–1110.
3. Amerini, Irene, Lamberto Ballan, Roberto Caldelli, Alberto Del Bimbo, Luca Del Tongo, and Giuseppe Serra. "Copy-move forgery detection and localization by means of robust clustering with J-Linkage." *Signal Processing: Image Communication* 28, no. 6 (2013): 659–669.

4. Ardizzone, Edoardo, Alessandro Bruno, and Giuseppe Mazzola. "Detecting multiple copies in tampered images." In 2010 IEEE International Conference on Image Processing, pp. 2117–2120. IEEE, 2010.
5. Bayram, Sevinc, Husrev Taha Sencar, and Nasir Memon. "An efficient and robust method for detecting copy-move forgery." In 2009 IEEE International Conference on Acoustics, Speech and Signal Processing, pp. 1053–1056. IEEE, 2009.
6. Bianchi, Tiziano, and Alessandro Piva. "Detection of non-aligned double JPEG compression with estimation of primary compression parameters." In 2011 18th IEEE International Conference on Image Processing, pp. 1929–1932. IEEE, 2011.
7. Birajdar, Gajanan K., and Vijay H. Mankar. "Digital image forgery detection using passive techniques: A survey." *Digital Investigation* 10, no. 3 (2013): 226–245.
8. Jaberi, Maryam, George Bebis, Muhammad Hussain, and Ghulam Muhammad. "Accurate and robust localization of duplicated region in copy–move image forgery." *Machine Vision and Applications* 25 (2014): 451–475.
9. Cao, Gang, Yao Zhao, Rongrong Ni, and Alex C. Kot. "Unsharp masking sharpening detection via overshoot artifacts analysis." *IEEE Signal Processing Letters* 18, no. 10 (2011): 603–606.
10. Cao, Gang, Yao Zhao, Rongrong Ni, and Xuelong Li. "Contrast enhancement-based forensics in digital images." *IEEE Transactions on Information Forensics and Security* 9, no. 3 (2014): 515–525.
11. Christlein, Vincent, Christian Riess, Johannes Jordan, Corinna Riess, and Elli Angelopoulou. "An evaluation of popular copy-move forgery detection approaches." *IEEE Transactions on Information Forensics and Security* 7, no. 6 (2012): 1841–1854.
12. Cozzolino, Davide, Giovanni Poggi, and Luisa Verdoliva. "Efficient dense-field copy–move forgery detection." *IEEE Transactions on Information Forensics and Security* 10, no. 11 (2015): 2284–2297.
13. Farid, H. Exposing digital forgeries from JPEG ghosts. *IEEE Transactions on Information Forensics and Security* 4, no. 1 (2009): 154–160.
14. Farid, Hany. "Image forgery detection." *IEEE Signal Processing Magazine* 26, no. 2 (2009): 16–25.
15. Ferrara, Pasquale, Tiziano Bianchi, Alessia De Rosa, and Alessandro Piva. "Reverse engineering of double compressed images in the presence of contrast enhancement." In 2013 IEEE 15th International Workshop on Multimedia Signal Processing (MMSP), pp. 141–146. IEEE, 2013.
16. Gharibi, Fereshteh, Javad Ravan Jamjah, Fardin Akhlaghian, Bahram Zahir Azami, and Javad Alirezaie. "Robust detection of copy-move forgery using texture features." In 2011 19th Iranian Conference on Electrical Engineering, pp. 1–4. IEEE, 2011.
17. Zhou, Guojuan, and Dianji Lv. "An overview of digital watermarking in image forensics." In 2011 Fourth International Joint Conference on Computational Sciences and Optimization, pp. 332–335. IEEE, 2011.
18. Hashmi, Mohammad Farukh, Aaditya R. Hambarde, and Avinash G. Keskar. "Copy move forgery detection using DWT and SIFT features." In 2013 13th International Conference on Intellient Systems Design and Applications, pp. 188–193. IEEE, 2013.
19. Hashmi, Mohammad Farukh, Vijay Anand, and Avinash G. Keskar. "A copy-move image forgery detection based on speeded up robust feature transform and Wavelet Transforms." In 2014 International Conference on Computer and Communication Technology (ICCCT), pp. 147–152. IEEE, 2014.
20. Hsu, Chen-Ming, Jen-Chun Lee, and Wei-Kuei Chen. "An efficient detection algorithm for copy-move forgery." In 2015 10th Asia Joint Conference on Information Security, pp. 33–36. IEEE, 2015.

21. Jaberi, Maryam, George Bebis, Muhammad Hussain, and Ghulam Muhammad. "Accurate and robust localization of duplicated region in copy–move image forgery." *Machine Vision and Applications* 25 (2014): 451–475.
22. Jing, Li, and Chao Shao. "Image copy-move forgery detecting based on local invariant feature." *Journal of Multimedia* 7, no. 1 (2012).
23. Khan, Saiqa, and Arun Kulkarni. "Robust method for detection of copy-move forgery in digital images." In 2010 International Conference on Signal and Image Processing, pp. 69–73. IEEE, 2010.
24. Li, Jian, Xiaolong Li, Bin Yang, and Xingming Sun. "Segmentation-based image copy-move forgery detection scheme." *IEEE Transactions on Information Forensics and Security* 10, no. 3 (2014): 507–518.
25. Li, Kunlun, Hexin Li, Bo Yang, Qi Meng, and Shangzong Luo. "Detection of image forgery based on improved PCA-SIFT." In Computer Engineering and Networking: Proceedings of the 2013 International Conference on Computer Engineering and Network (CENet2013), pp. 679–686. Springer International Publishing, 2014.
26. Li, Leida, Shushang Li, Hancheng Zhu, and Shu-Chuan Chu. An efficient scheme for detecting copy-move forged images by local binary patterns. *J. Inf. Hiding Multim. Signal Process,* 4, no. 1 (2013): 46–56.
27. Li, Weihai, and Nenghai Yu. "Rotation robust detection of copy-move forgery." In 2010 IEEE International Conference on Image Processing, pp. 2113–2116. IEEE, 2010.
28. Mahalakshmi, S. Devi, K. Vijayalakshmi, and S. Priyadharsini. "Digital image forgery detection and estimation by exploring basic image manipulations." *Digital Investigation* 8, no. 3–4 (2012): 215–225.
29. Muhammad, Najah, Muhammad Hussain, Ghulam Muhammad, and George Bebis. "Copy-move forgery detection using dyadic wavelet transform." In 2011 Eighth International Conference Computer Graphics, Imaging and Visualization, pp. 103–108. IEEE, 2011.
30. Panchal, P.M., S.R. Panchal, and S.K. Shah. "A comparison of SIFT and SURF." *International Journal of Innovative Research in Computer and Communication Engineering* 1, no. 2 (2013): 323–327.
31. Pan, Xunyu, and Siwei Lyu. "Region duplication detection using image feature matching." *IEEE Transactions on Information Forensics and Security* 5, no. 4 (2010): 857–867.
32. Qian, Ruohan, Weihai Li, Nenghai Yu, and Zhuo Hao. "Image forensics with rotation-tolerant resampling detection." In 2012 IEEE International Conference on Multimedia and Expo Workshops, pp. 61–66. IEEE, 2012.
33. Redi, Judith A., Wiem Taktak, and Jean-Luc Dugelay. "Digital image forensics: A booklet for beginners." *Multimedia Tools and Applications* 51 (2011): 133–162.
34. Sridevi, M., C. Mala, and Siddhant Sanyam. "Comparative study of image forgery and copy-move techniques." In Advances in Computer Science, Engineering & Applications: Proceedings of the Second International Conference on Computer Science, Engineering and Applications (ICCSEA 2012), May 25-27, 2012, New Delhi, India, Volume 1, pp. 715–723. Springer Berlin Heidelberg, 2012.
35. Sunil, Kumar, Desai Jagan, and Mukherjee Shaktidev. "DCT-PCA based method for copy-move forgery detection." In ICT and Critical Infrastructure: Proceedings of the 48th Annual Convention of Computer Society of India-Vol II: Hosted by CSI Vishakapatnam Chapter, pp. 577–583. Springer International Publishing, 2014.

36. Garg, Ankit, Aleem Ali, and Puneet Kumar. "Original Research Article A shadow preservation framework for effective content-aware image retargeting process." *Journal of Autonomous Intelligence* 6, no. 3 (2023): 282–296.
37. Wang, Junbin, Zhenghong Yang, and Shaozhang Niu. "Copy-move forgeries detection based on SIFT algorithm." *Proceedings of the International Journal of Computer Science* 2, (2015): 567–570.
38. Yu, Liyang, Qi Han, and Xiamu Niu. "Feature point-based copy-move forgery detection: Covering the non-textured areas." *Multimedia Tools and Applications* 75 (2016): 1159–1176.
39. Kumar, Sunil, and P. K. Das. "Copy-move forgery detection in digital images: Progress and challenges." *International Journal on Computer Science and Engineering* 3, no. 2 (2011): 652–663.
40. Hamid, Irfan, Rameez Raja, Monika Anand, Vijay Karnatak, and Aleem Ali. "Comprehensive robustness evaluation of an automatic writer identification system using convolutional neural networks." (2023).
41. Sachdeva, Shaweta, and Aleem Ali. "Machine learning with digital forensics for attack classification in cloud network environment." *International Journal of System Assurance Engineering and Management* 13, no. Suppl 1 (2022): 156–165.
42. Ansari, Farah Jamal, and Aleem Ali. "A comparison of the DCT JPEG and wavelet image compression encoder for medical images." In Contemporary Computing: 5th International Conference, IC3 2012, Noida, India, August 6-8, 2012. Proceedings 5, pp. 490–491. Springer Berlin Heidelberg, 2012.
43. Sachdeva, Shaweta, and Aleem Ali. "A hybrid approach using digital Forensics for attack detection in a cloud network environment." *International Journal of Future Generation Communication and Networking* 14, no. 1 (2021): 1536–1546.

3 Artificial Intelligence (AI) and Optimization for Health Information System (HIS) Generation

Payal Thakur, Shanu Khare, and Navjot Singh Talwandi

3.1 INTRODUCTION

3.1.1 Defining the Landscape: AI and Optimization in Health Information Systems

AI and optimization play a crucial role in health information systems, revolutionizing the way healthcare organizations manage and utilize data[1]. These technologies enable healthcare providers to improve patient care, enhance operational efficiency, and make data-driven decisions. AI involves the development of intelligent machines that can perform tasks that typically require human intelligence. In health information systems, AI is used to analyze vast amounts of patient data, identify patterns, and make predictions or recommendations. Some key applications of AI in healthcare include:

Diagnosis and Treatment: AI algorithms can analyze medical images, such as X-rays or MRIs, to detect abnormalities or assist in diagnosing diseases. They can also suggest treatment plans based on patient data and medical guidelines[2].

Predictive Analytics: AI can analyze patient data, including medical history, lab results, and vital signs, to predict disease progression, identify high-risk patients, and recommend preventive measures.

Natural Language Processing (NLP): NLP enables computers to understand and interpret human language. In health information systems, NLP can be used to extract relevant information from medical records, automate coding, and improve clinical documentation.

Virtual Assistants: AI-powered virtual assistants, like chatbots, can provide patients with basic medical information, answer common questions, and schedule appointments. They can also assist healthcare professionals by providing quick access to medical knowledge and guidelines.

Optimization in Health Information Systems Optimization techniques aim to find the best possible solution to a problem within given constraints[3]. In health information systems, optimization is used to improve various aspects of healthcare operations, such as:

Resource Allocation: Optimization models can help healthcare organizations allocate resources, such as staff, equipment, and beds, efficiently. These models consider factors like patient demand, resource availability, and operational constraints to optimize resource utilization.

Scheduling and Routing: Optimization algorithms can optimize appointment scheduling, surgery scheduling, and patient routing within a healthcare facility. These algorithms consider factors like patient preferences, healthcare provider availability, and travel distances to minimize waiting times and maximize efficiency.

Supply Chain Management: Optimization techniques can optimize inventory management, procurement, and distribution of medical supplies and medications. By considering factors like demand variability, lead times, and storage costs, these models can minimize stockouts and reduce costs.

Workflow Optimization: Optimization models can analyze and optimize workflows within healthcare organizations, identifying bottlenecks, streamlining processes, and improving overall efficiency.

By leveraging AI and optimization techniques, health information systems can enhance patient care, improve operational efficiency, and enable data-driven decision-making in healthcare organizations[4]. These technologies have the potential to transform the healthcare industry and improve outcomes for patients.

3.1.2 The Evolution of Health Information Systems

Health information systems have evolved significantly over the years, driven by advancements in technology and the increasing need for efficient and effective healthcare delivery. In the early days, healthcare organizations relied on paper-based systems for storing and managing patient information. This involved manual record-keeping, filing, and retrieval of medical records[5]. While this method was widely used, it was time-consuming, prone to errors, and limited in terms of data accessibility. The introduction of electronic health records revolutionized health information systems. EHRs digitized patient records, making them easily accessible, searchable, and shareable across healthcare providers. EHRs improved data accuracy, reduced paperwork, and facilitated better coordination of care among different healthcare professionals. As healthcare organizations adopted EHRs, the need for interoperability and health information exchange became apparent[6]. Interoperability refers to the ability of different systems to exchange and use data seamlessly. Health information exchange enables the secure sharing of patient information between healthcare providers, improving care coordination and continuity.

Decision support systems (DSS) integrated with EHRs provide healthcare professionals with real-time clinical decision support. DSS uses algorithms and data analysis to provide evidence-based recommendations, alerts, and reminders to healthcare providers, improving patient safety and clinical outcomes[7]. The advancement of technology has enabled the growth of telemedicine and remote monitoring systems. These systems allow healthcare providers to remotely monitor patients, conduct virtual consultations, and provide care outside traditional healthcare settings. Telemedicine and remote monitoring systems have become particularly important during the COVID-19 pandemic, enabling safe and accessible healthcare delivery. AI and analytics have emerged as powerful tools in health information systems. AI algorithms can analyze large volumes of patient data, identify patterns, and make predictions or recommendations. Analytics tools enable healthcare organizations to gain insights from data, identify trends, and make data-driven decisions to improve patient care and operational efficiency. The proliferation of smartphones and wearable devices has led to the growth of mobile health (mHealth) applications. These applications allow patients to monitor their health, access medical information, and communicate with healthcare providers. Wearable devices, such as fitness trackers and smartwatches, collect real-time health data, which can be integrated into health information systems for personalized care and remote monitoring[8].

The evolution of health information systems has transformed the way healthcare is delivered, improving patient care, enhancing data accessibility, and enabling better decision-making. As technology continues to advance, health information systems will likely continue to evolve, incorporating innovations like AI, machine learning, and blockchain to further enhance healthcare delivery[9].

3.2 FOUNDATIONS OF HEALTH INFORMATION SYSTEMS

3.2.1 Key Components and Architecture

Health information systems (HIS) consist of various components and follow a specific architecture to ensure efficient management and utilization of healthcare data[10]. Here are the key components and architecture of health information systems:

Data Sources: Health information systems gather data from various sources, including electronic health records (EHRs), laboratory systems, medical imaging systems, pharmacy systems, and wearable devices. These sources generate and store patient-related data, such as demographics, medical history, test results, and treatment plans.

Data Storage: The data collected from different sources is stored in a centralized database or data warehouse. This storage system ensures data integrity, security, and accessibility. It allows for efficient data retrieval and supports data analytics and reporting.

Data Integration: Health information systems integrate data from different sources to create a comprehensive patient record. This integration ensures that all relevant information is available to healthcare providers for decision-making and care coordination. Data integration may involve standardization and mapping of data elements to ensure consistency and interoperability[11].

Data Exchange: Health information systems facilitate the exchange of patient data between different healthcare organizations and systems. This enables seamless sharing of information for care coordination, referrals, and transitions of care. Standards like HL7 (Health Level Seven) and FHIR (Fast Healthcare Interoperability Resources) are used to ensure interoperability and data exchange.

Clinical Decision Support: Health information systems incorporate clinical decision support systems (CDSS) to provide healthcare professionals with evidence-based recommendations, alerts, and reminders. CDSS uses algorithms and medical knowledge databases to assist in diagnosis, treatment planning, and medication management. It helps improve patient safety, adherence to guidelines, and clinical outcomes.

User Interface: Health information systems have user interfaces that allow healthcare professionals to interact with the system. These interfaces can be web-based or desktop applications and provide functionalities like data entry, retrieval, visualization, and reporting. User interfaces are designed to be intuitive, user-friendly, and tailored to the specific needs of different healthcare roles[12].

Security and Privacy: Health information systems prioritize security and privacy to protect patient data. They implement measures like access controls, encryption, audit trails, and user authentication to ensure data confidentiality and integrity. Compliance with regulations like HIPAA (Health Insurance Portability and Accountability Act) is essential to safeguard patient information.

Analytics and Reporting: Health information systems incorporate analytics tools to analyze and derive insights from healthcare data. These tools enable data mining, trend analysis, and reporting to support quality improvement initiatives, population health management, and research. Analytics capabilities help identify patterns, measure performance, and support data-driven decision-making.

The architecture of health information systems can vary depending on the specific implementation and the organization's requirements as shown in Figure 3.1. It

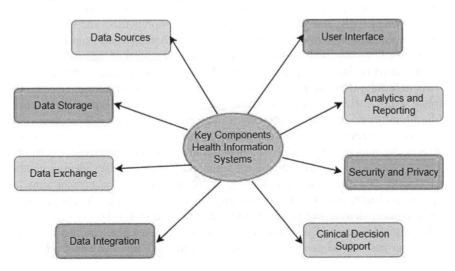

FIGURE 3.1 Key Components of Health information systems (HIS).

may involve a combination of on-premises servers, cloud-based infrastructure, and interoperability standards to ensure seamless data flow and system integration[13]. Overall, health information systems aim to centralize and manage healthcare data efficiently, support clinical decision-making, improve patient care, and enable data-driven healthcare delivery.

3.2.2 Historical Perspectives on HIS

Before the advent of computers, healthcare organizations relied on paper-based systems for managing patient information. This involved manual record-keeping, filing, and retrieval of medical records. While this method was widely used, it was time-consuming, prone to errors, and limited in terms of data accessibility. The development of computers and information technology in the 1960s and 1970s paved the way for the digitization of health records. The first electronic health record systems were introduced in the 1970s, enabling healthcare organizations to store and manage patient information electronically. These early EHRs were often limited to specific departments or functions within healthcare organizations[14].

In the 1980s and 1990s, efforts were made to standardize health information systems and promote interoperability. Standards like HL7 (Health Level Seven) were developed to facilitate the exchange of health information between different systems and organizations. The goal was to ensure that healthcare data could be shared and used across different healthcare settings. In the late 1990s and early 2000s, the concept of health information exchanges (HIEs) gained prominence. HIEs aimed to facilitate the secure sharing of patient information between healthcare providers, enabling better care coordination and continuity. HIEs played a crucial role in promoting interoperability and data exchange between different health information systems. In 2009, the U.S. government introduced the Meaningful Use program as part of the Health Information Technology for Economic and Clinical Health (HITECH) Act. This program provided financial incentives to healthcare organizations that adopted and demonstrated meaningful use of certified EHR systems. This initiative accelerated the adoption of EHRs and paved the way for more comprehensive and standardized health information systems. The advancement of technology, particularly in areas like cloud computing, mobile devices, and artificial intelligence, has had a significant impact on health information systems. Cloud-based systems have enabled easier access to health information, improved scalability, and reduced infrastructure costs. Mobile health (mHealth) applications and wearable devices have expanded the reach of health information systems, allowing patients to actively participate in their own care. In recent years, there has been a growing emphasis on data analytics and population health management within health information systems. Analytics tools and techniques are being used to analyze large volumes of healthcare data, identify trends, and support data-driven decision-making. Population health management aims to improve the health outcomes of specific populations by leveraging health information systems to identify and address health risks and disparities[15].

The historical perspectives on health information systems highlight the evolution from paper-based systems to sophisticated electronic systems that enable efficient data management, interoperability, and data-driven healthcare delivery. As technology continues to advance, health information systems will likely continue to evolve, incorporating innovations like artificial intelligence, machine learning, and blockchain to further enhance healthcare delivery and outcomes.

3.3 AI FUNDAMENTALS

3.3.1 BASICS OF ARTIFICIAL INTELLIGENCE

Artificial Intelligence (AI) refers to the development of intelligent machines that can perform tasks that typically require human intelligence. AI systems are designed to perceive their environment, reason, learn from experience, and make decisions or take actions to achieve specific goals.

Machine learning is a subset of AI that focuses on enabling machines to learn from data and improve their performance without being explicitly programmed. Machine learning algorithms can analyze large datasets, identify patterns, and make predictions or decisions based on the learned patterns. Common machine learning techniques include supervised learning, unsupervised learning, and reinforcement learning. Deep learning is a subfield of machine learning that uses artificial neural networks to model and understand complex patterns and relationships in data. Deep learning algorithms are inspired by the structure and function of the human brain, with multiple layers of interconnected nodes (neurons) that process and transform data. Deep learning has achieved remarkable success in areas like image recognition, natural language processing, and speech recognition. NLP enables computers to understand, interpret, and generate human language. It involves tasks like speech recognition, language translation, sentiment analysis, and text generation. NLP algorithms use techniques like text parsing, semantic analysis, and machine translation to process and understand human language[16].

Computer vision is a field of AI that focuses on enabling machines to understand and interpret visual information from images or videos. Computer vision algorithms can analyze and extract features from images, recognize objects or patterns, and perform tasks like object detection, image classification, and image segmentation. Computer vision has applications in areas like autonomous vehicles, surveillance systems, and medical imaging. Robotics combines AI with mechanical engineering to create intelligent machines or robots that can interact with the physical world. AI-powered robots can perceive their environment, make decisions, and perform physical tasks. Robotics has applications in various industries, including manufacturing, healthcare, and agriculture. Expert systems are AI systems that emulate the knowledge and decision-making capabilities of human experts in specific domains. These systems use rules and logic to solve complex problems and provide expert-level recommendations or solutions. Expert systems have been used in areas like medical diagnosis, financial analysis, and customer support. As AI continues to advance, ethical considerations become increasingly important. Issues like bias in AI algorithms, privacy concerns, and the impact of AI on jobs and society need to be addressed. Responsible AI development involves ensuring fairness, transparency, and accountability in AI systems.

3.3.2 MACHINE LEARNING IN HEALTHCARE

Machine learning (ML) has emerged as a powerful tool in healthcare, revolutionizing various aspects of the industry. Here are some key applications of machine learning in healthcare:

Medical Imaging: ML algorithms can analyze medical images, such as X-rays, MRIs, and CT scans, to detect abnormalities, assist in diagnosis, and predict. disease progression. ML models can learn from large datasets of labeled images to identify patterns and make accurate predictions, helping radiologists and clinicians in their decision-making process.

Disease Diagnosis and Risk Prediction: ML algorithms can analyze patient data, including medical history, lab results, and genetic information, to assist in disease diagnosis and predict the risk of developing certain conditions. ML models can identify patterns and risk factors that may not be apparent to human experts, enabling early detection and personalized treatment plans.

Drug Discovery and Development: ML is being used to accelerate the drug discovery and development process. ML models can analyze large datasets of molecular structures, genetic information, and clinical trial data to identify potential drug candidates, predict their efficacy, and optimize drug design. This can help reduce the time and cost involved in bringing new drugs to market.

Personalized Medicine: ML algorithms can analyze patient data, including genetic information and treatment outcomes, to develop personalized treatment plans. ML models can identify patient-specific factors that influence treatment response and recommend the most effective interventions. This can lead to improved patient outcomes and reduced healthcare costs.

Electronic Health Records (EHR) Analysis: ML can analyze large volumes of EHR data to extract valuable insights and support clinical decision-making. ML models can identify patterns in patient data, predict disease progression, and recommend appropriate treatments. ML can also automate tasks like coding, documentation, and anomaly detection in EHRs, improving efficiency and accuracy[17].

Remote Patient Monitoring: ML algorithms can analyze data from wearable devices and remote monitoring systems to track patient health, detect anomalies, and provide real-time alerts. ML models can identify patterns in physiological data, such as heart rate, blood pressure, and glucose levels, to monitor chronic conditions and enable early intervention.

Healthcare Operations and Resource Management: ML can optimize healthcare operations by analyzing data related to patient flow, resource utilization, and scheduling. ML models can predict patient demand, optimize appointment scheduling, and allocate resources efficiently, leading to improved operational efficiency and reduced wait times.

It is important to note that the successful implementation of machine learning in healthcare requires high-quality and diverse datasets, robust algorithms, and careful validation and integration into clinical workflows. Additionally, privacy and security considerations must be addressed to protect patient data. Nonetheless, machine learning holds great promise in transforming healthcare delivery, improving patient outcomes, and advancing medical research.

3.4 OPTIMIZATION TECHNIQUES FOR HEALTH INFORMATION SYSTEMS (HIS)

3.4.1 Principles of Optimization

Optimization principles play a crucial role in designing and managing health information systems (HIS) to ensure efficient and effective healthcare delivery. Clearly define the objectives and goals of the health information system. This includes identifying the specific outcomes or improvements that the system aims to achieve, such as enhancing data accuracy, improving care coordination, or increasing operational efficiency. Determine the KPIs that will be used to measure the performance and success of the health information system. KPIs may include metrics like data accuracy rates, turnaround times, patient satisfaction scores, or cost savings. These indicators help assess the effectiveness of the system and identify areas for improvement[18]. Optimize workflows within the health information system to minimize redundancies, eliminate unnecessary steps, and improve efficiency. This involves mapping out the flow of information and processes, identifying bottlenecks or inefficiencies, and implementing changes to streamline operations. Data quality is crucial for the effectiveness of health information systems. Implement measures to ensure data accuracy, completeness, consistency, and timeliness. This may involve data validation checks, data cleansing processes, and regular data audits. High-quality data is essential for informed decision-making and reliable reporting. Interoperability is the ability of different systems and applications to exchange and use data seamlessly. Ensure that the health information system is designed to integrate and communicate with other systems, such as electronic health records, laboratory systems, and pharmacy systems. This enables the sharing of patient information and promotes care coordination. Efficient data storage and retrieval are critical for health information systems. Implement appropriate data storage solutions, such as databases or data warehouses that can handle large volumes of data and provide fast and reliable access. Indexing and search functionalities should be optimized to enable quick and accurate retrieval of information.

3.4.2 Application of Optimization in Health Information Systems

Optimization techniques can be applied to various aspects of health information systems (HIS) to improve their efficiency, effectiveness, and overall performance. Here are some key areas where optimization can be applied in health information systems:

Data Management: Optimization techniques can be used to improve data management processes within the HIS. This includes optimizing data storage and retrieval mechanisms, ensuring data quality and integrity, and implementing efficient data cleansing and validation processes. By optimizing data management, the HIS can provide accurate and reliable information for decision-making and reporting.

Workflow and Process Optimization: Optimization can be applied to streamline workflows and processes within the HIS. This involves identifying bottlenecks, eliminating redundancies, and automating manual tasks. By optimizing workflows, the HIS can improve operational efficiency, reduce turnaround times, and enhance overall productivity.

Resource Allocation: Optimization techniques can be used to optimize resource allocation within the HIS. This includes optimizing the allocation of hardware resources, such as servers and storage, to ensure optimal performance and scalability. Additionally, optimization can be applied to allocate human resources effectively, such as assigning tasks and responsibilities based on skill sets and workload.

System Integration and Interoperability: Optimization can be applied to ensure seamless integration and interoperability between different systems and applications within the HIS. This involves optimizing data exchange protocols, standardizing data formats, and implementing efficient data mapping and transformation processes. By optimizing system integration and interoperability, the HIS can facilitate the seamless flow of information across different healthcare settings and improve care coordination.

Decision Support Systems: Optimization techniques can be applied to develop decision support systems within the HIS. This involves using algorithms and models to analyze data, identify patterns, and provide recommendations for clinical decision-making. By optimizing decision support systems, the HIS can assist healthcare providers in making informed decisions, improving patient outcomes, and reducing medical errors.

Performance Monitoring and Analytics: Optimization techniques can be used to monitor and analyze the performance of the HIS. This includes implementing performance monitoring tools, analyzing system logs and metrics, and identifying areas for improvement. By optimizing performance monitoring and analytics, the HIS can identify bottlenecks, optimize system configurations, and proactively address issues to ensure optimal system performance.

Security and Privacy: Optimization techniques can be applied to enhance the security and privacy of the HIS. This includes optimizing access controls, encryption algorithms, and authentication mechanisms to protect patient data from unauthorized access or breaches. By optimizing security and privacy measures, the HIS can ensure compliance with regulations and build trust among users.

Overall, optimization plays a crucial role in improving the efficiency, effectiveness, and performance of health information systems. By applying optimization techniques to various aspects of the HIS, healthcare organizations can enhance data management, streamline workflows, improve resource allocation, facilitate interoperability, support decision-making, monitor performance, and ensure the security and privacy of patient data.

3.5 INTEGRATION OF AI IN HEALTH INFORMATION SYSTEMS

3.5.1 ENHANCING DATA PROCESSING WITH AI

Artificial Intelligence (AI) can greatly enhance data processing capabilities by automating and optimizing various tasks. AI algorithms can automate the process of cleaning and preprocessing data. AI models can identify and handle missing values, outliers, and inconsistencies in the data, improving data quality and accuracy. This saves time and effort compared to manual data cleaning processes. AI can help integrate and fuse data from multiple sources. AI algorithms can analyze and match data from different formats and structures, enabling the integration of diverse datasets. This allows for a comprehensive view of the data and facilitates more accurate analysis and decision-making[19]. AI techniques, such as machine learning, can automatically classify and categorize data based on patterns and features. This can be useful for organizing and structuring large datasets, making it easier to search, retrieve, and analyze specific subsets of data. NLP techniques enable AI systems to understand and process human language. AI-powered NLP algorithms can extract information, sentiment, and context from unstructured text data, such as medical records, research papers, or social media posts. This enables efficient analysis and extraction of valuable insights from textual data. AI algorithms, particularly machine learning and deep learning models, can analyze large volumes of data to identify patterns, correlations, and trends. This allows for more accurate predictions, anomaly detection, and data-driven decision-making. AI can uncover hidden insights and relationships that may not be apparent to human analysts.

3.5.2 AI-DRIVEN DECISION SUPPORT SYSTEMS IN HEALTHCARE

AI-driven decision support systems (DSS) in healthcare leverage artificial intelligence techniques to assist healthcare professionals in making informed and evidence-based decisions. Here are some key aspects of AI-driven DSS in healthcare:

Clinical Decision Support: AI-driven DSS can provide healthcare professionals with real-time, personalized recommendations and alerts based on patient data, medical guidelines, and research evidence. These systems can assist in diagnosis, treatment selection, medication management, and monitoring of patient conditions. By analyzing large volumes of data and applying machine learning algorithms, AI-driven DSS can improve clinical decision-making and enhance patient outcomes.

Risk Prediction and Stratification: AI-driven DSS can analyze patient data, including medical history, lab results, and genetic information, to predict the risk of developing certain diseases or adverse events. These systems can stratify patients into risk categories, enabling targeted interventions and preventive measures. AI models can identify patterns and risk factors that may not be apparent to human experts, assisting in early detection and proactive management of health conditions.

Precision Medicine: AI-driven DSS can support precision medicine approaches by analyzing patient-specific data, such as genetic information, biomarkers, and treatment outcomes. These systems can identify patient subgroups that are likely to respond to specific treatments or therapies, enabling personalized treatment plans. AI models can integrate diverse data sources and provide recommendations for tailored interventions, optimizing patient care.

Image and Signal Analysis: AI-driven DSS can analyze medical images, such as X-rays, MRIs, and CT scans, to assist in diagnosis and treatment planning. These systems can detect abnormalities, segment organs or lesions, and provide quantitative measurements. AI models, particularly deep learning algorithms, can learn from large datasets of labeled images to improve accuracy and efficiency in image interpretation, supporting radiologists and clinicians.

Natural Language Processing (NLP): AI-driven DSS can leverage NLP techniques to extract and analyze information from unstructured clinical notes, research articles, and other textual sources. These systems can assist in literature review, evidence synthesis, and clinical documentation. NLP algorithms can identify relevant concepts, relationships, and trends, providing healthcare professionals with timely and comprehensive information.

Adverse Event Detection and Surveillance: AI-driven DSS can monitor and analyze healthcare data to detect adverse events, medication errors, and other safety concerns. These systems can identify patterns, outliers, and signals of potential harm, enabling proactive interventions and quality improvement initiatives. AI models can analyze structured and unstructured data sources, including electronic health records and patient-reported data, to enhance patient safety.

Workflow Optimization: AI-driven DSS can optimize healthcare workflows by automating routine tasks, prioritizing worklists, and providing decision support at the point of care. These systems can integrate with electronic health record systems and other clinical tools to streamline processes and reduce administrative burden. AI models can learn from user interactions and feedback to continuously improve system performance and usability.

AI-driven DSS in healthcare have the potential to improve clinical outcomes, enhance patient safety, and optimize healthcare delivery. However, it is important to ensure the transparency, explainability, and ethical use of AI models in decision-making processes. Collaboration between healthcare professionals, data scientists, and regulatory bodies is crucial to develop and deploy AI-driven DSS that are reliable, trustworthy, and aligned with clinical needs.

3.6 DATA MANAGEMENT AND GOVERNANCE

3.6.1 Data Quality and Integrity in HIS

Data quality and integrity are critical aspects of health information systems (HIS) to ensure accurate and reliable information for healthcare delivery and decision-making. Here are some key considerations for maintaining data quality and integrity in HIS:

Data Governance: Establish a robust data governance framework that defines roles, responsibilities, and processes for data management. This includes clear data ownership, data stewardship, and data quality management practices. Data governance ensures accountability and promotes a culture of data quality and integrity.

Data Standardization: Implement standardized data models, coding systems, and terminologies to ensure consistency and interoperability of data across different systems and healthcare settings. Standardization facilitates data exchange, aggregation, and analysis, reducing errors and inconsistencies in data.

Data Validation and Verification: Implement data validation checks and verification processes to ensure the accuracy, completeness, and consistency of data. This includes validating data against predefined rules, conducting data audits, and reconciling data with source documents or systems. Regular data quality assessments help identify and rectify data errors or discrepancies.

Data Entry and Documentation Standards: Establish clear data entry and documentation standards to ensure consistent and accurate data capture. This includes defining data fields, formats, and guidelines for data entry. Training and education programs should be provided to healthcare professionals to ensure adherence to data entry standards.

Data Privacy and Security: Implement robust data privacy and security measures to protect the confidentiality, integrity, and availability of data. This includes access controls, encryption, user authentication, and audit trails. Compliance with relevant regulations, such as HIPAA (Health Insurance Portability and Accountability Act), is essential to safeguard patient data.

Data Integration and Interoperability: Ensure seamless integration and interoperability of data across different systems and applications within the HIS. This involves implementing standardized data exchange protocols, data mapping, and transformation processes. Data integration and interoperability reduce data duplication, inconsistencies, and errors.

Data Quality Monitoring and Reporting: Establish mechanisms to monitor and report data quality metrics and issues. This includes implementing data quality dashboards, generating data quality reports, and conducting regular data quality audits. Monitoring data quality helps identify trends, patterns, and areas for improvement.

Data Cleansing and Enrichment: Implement data cleansing processes to identify and correct errors, inconsistencies, and duplicates in the data. Data enrichment techniques, such as data linking or data augmentation, can be used to enhance the completeness and accuracy of data. These processes improve data quality and integrity.

User Training and Education: Provide comprehensive training and education programs to healthcare professionals and system users on data quality principles, data entry standards, and best practices. User awareness and understanding of the importance of data quality contribute to maintaining data integrity.

Continuous Improvement: Establish a culture of continuous improvement in data quality and integrity. This involves regularly reviewing data quality processes, addressing identified issues, and implementing corrective actions. Feedback mechanisms from users and stakeholders should be incorporated to drive ongoing enhancements.

By implementing these measures, healthcare organizations can ensure data quality and integrity in their health information systems. High-quality and reliable data support evidence-based decision-making, improve patient care, and enable effective healthcare management.

3.6.2 GOVERNANCE FRAMEWORKS FOR AI-ENHANCED SYSTEMS

Governance frameworks for AI-enhanced systems involve establishing guidelines, policies, and ethical principles to regulate the development, deployment, and use of artificial intelligence in various domains. In the context of AI-enhanced systems, particularly in healthcare or other critical sectors, robust governance frameworks are essential to ensure accountability, transparency, fairness, and ethical use of AI technologies. Clear ethical guidelines define the principles and values that AI systems should adhere to, ensuring they prioritize human wellbeing, fairness, privacy, and accountability. These guidelines often stem from established ethical frameworks and human rights principles[19]. Compliance with existing regulations and standards is crucial. New regulations specific to AI might also be necessary to address unique challenges posed by these technologies, such as data privacy, algorithm transparency, and liability issues. Establishing mechanisms for accountability ensures that responsibilities are defined for various stakeholders involved in AI development, deployment, and usage. Transparency requirements mandate that AI systems provide understandable explanations for their decisions or actions.

Assessing potential risks associated with AI systems, such as biases, errors, or unintended consequences, and implementing strategies to mitigate these risks is integral. This includes continuous monitoring and evaluation of AI systems. Robust measures for data protection and security are vital to safeguard sensitive information used or generated by AI systems. Compliance with privacy regulations and best practices in data handling is a fundamental aspect. Involving stakeholders from various domains, including policymakers, industry experts, ethicists, and the public, fosters diverse perspectives and ensures the development of more comprehensive governance frameworks. Providing education and training on AI ethics and governance to stakeholders, including developers, users, and policymakers, is essential for understanding the implications and responsibilities associated with AI technologies[20].

Implementing effective governance frameworks requires collaboration between governments, industry leaders, researchers, and the public to create balanced, adaptable, and responsive regulations that promote the beneficial use of AI while mitigating potential risks and ethical concerns.

3.7 MACHINE LEARNING APPLICATIONS IN HEALTHCARE

3.7.1 Predictive Modeling for Disease Management

Predictive modeling for disease management involves using AI and machine learning techniques to analyze patient data and make predictions about disease outcomes, treatment effectiveness, and patient risk factors. Here are some key considerations for implementing predictive modeling in disease management:

- **Data Collection and Integration:** Collect and integrate relevant patient data from various sources, such as electronic health records, medical imaging, wearable devices, and genetic information. Ensure data quality and standardization to enable accurate and reliable predictions.
- **Feature Selection and Engineering:** Identify the most relevant features or variables that contribute to disease prediction. This involves selecting appropriate clinical, demographic, and genetic factors and engineering new features that capture important information for the predictive model.
- **Model Development and Validation:** Develop predictive models using machine learning algorithms, such as logistic regression, decision trees, random forests, or deep learning models. Train and validate the models using appropriate datasets, ensuring robustness, accuracy, and generalizability.
- **Ethical Considerations:** Ensure that the use of predictive modeling in disease management adheres to ethical guidelines and principles. Protect patient privacy and confidentiality, obtain informed consent, and address potential biases or discrimination in the predictive models.
- **Interpretability and Explainability:** Aim for models that are interpretable and provide explanations for their predictions. This helps healthcare professionals understand the reasoning behind the predictions and gain trust in the model's recommendations.
- **Integration with Clinical Workflow:** Integrate predictive models into the clinical workflow to support healthcare professionals in decision-making. This may involve developing user-friendly interfaces, integrating with electronic health record systems, and providing real-time predictions and alerts.
- **Continuous Model Monitoring and Updating:** Continuously monitor the performance of predictive models and update them as new data becomes available. This ensures that the models remain accurate and relevant over time and adapt to changes in patient populations or treatment guidelines.
- **Validation and Clinical Trials:** Validate the predictive models through rigorous clinical trials and studies to assess their effectiveness and impact on patient outcomes. This helps establish the clinical utility and value of the predictive models in disease management.
- **Regulatory Compliance:** Ensure compliance with relevant regulations, such as data protection and privacy laws, when collecting, storing, and analyzing patient data for predictive modeling. Adhere to ethical guidelines and obtain necessary approvals from institutional review boards or ethics committees.

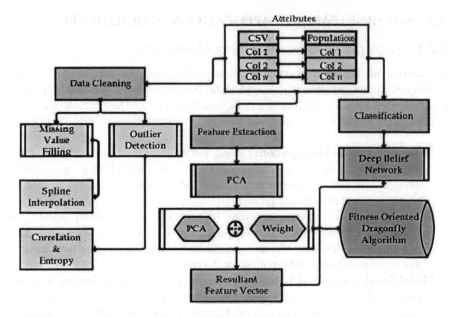

FIGURE 3.2 Predictive Modeling for Disease Management.

Collaboration and Stakeholder Engagement: Foster collaboration between data scientists, healthcare professionals, researchers, and patients to ensure the development and implementation of predictive modeling aligns with clinical needs and patient preferences. Engage stakeholders in the design, evaluation, and improvement of predictive models for disease management. Predictive Modeling for Disease Management in Figure 3.2.

By considering these factors, predictive modeling can be effectively applied in disease management to improve patient outcomes, optimize treatment strategies, and support healthcare decision-making.

3.7.2 Image and Speech Recognition in Health Information Systems

Image and speech recognition technologies have significant applications in health information systems (HIS) to enhance data processing and improve healthcare delivery. Image recognition algorithms can analyze medical images, such as X-rays, MRIs, CT scans, and pathology slides, to assist in diagnosis and treatment planning. These algorithms can detect abnormalities, segment organs or lesions, and provide quantitative measurements. Image recognition can help radiologists and clinicians in interpreting images more accurately and efficiently. Speech recognition technology can enable efficient and accurate transcription of spoken conversations between healthcare professionals and patients during telemedicine consultations. This allows for real-time documentation of patient encounters and facilitates remote collaboration and decision-making[21].

AI and Optimization for Health Information System Generation 63

Speech recognition can be used to convert spoken dictation into text, enabling healthcare professionals to create clinical documentation more efficiently. This reduces the time spent on manual data entry and allows for more accurate and timely documentation of patient encounters. Voice recognition technology can power virtual assistants that can respond to voice commands and provide information or perform tasks related to healthcare. These virtual assistants can assist healthcare professionals in accessing patient information, retrieving medical knowledge, and managing administrative tasks. NLP techniques can be combined with speech recognition to extract and analyze information from spoken conversations or transcribed text. NLP algorithms can identify relevant concepts, relationships, and sentiments, enabling advanced analysis of patient data and facilitating decision support. Image recognition can be used in wearable devices or remote monitoring systems to analyze images or videos captured by patients. For example, image recognition algorithms can analyze skin lesions or wounds to monitor healing progress or detect changes that may require medical attention. Image and speech recognition technologies can be used for quality assurance and compliance purposes. For example, image recognition algorithms can analyze medical images to ensure adherence to imaging protocols or identify potential errors. Speech recognition can be used to monitor and analyze conversations for compliance with regulatory requirements or quality standards. Speech recognition can be used to develop assistive technologies for individuals with disabilities or limited mobility. For example, speech recognition can enable hands-free control of medical devices or systems, allowing individuals to interact with HIS more easily (Figure 3.3).

Implementing image and speech recognition technologies in HIS requires careful consideration of data privacy, security, and regulatory compliance. It is important to ensure the accuracy, reliability, and ethical use of these technologies to maintain patient confidentiality and trust. Collaboration between healthcare professionals,

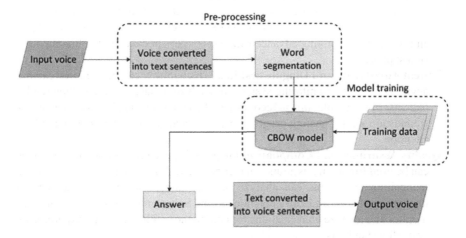

FIGURE 3.3 Image and Speech Recognition Technology.

data scientists, and technology providers is crucial to develop and deploy image and speech recognition solutions that meet the specific needs of healthcare settings.

3.8 ETHICAL CONSIDERATIONS IN AI FOR HEALTHCARE

Image and speech recognition technologies have significant applications in health information systems (HIS) to enhance data processing and improve healthcare delivery.

Medical Imaging Analysis: Image recognition algorithms can analyze medical images, such as X-rays, MRIs, CT scans, and pathology slides, to assist in diagnosis and treatment planning. These algorithms can detect abnormalities, segment organs or lesions, and provide quantitative measurements. Image recognition can help radiologists and clinicians in interpreting images more accurately and efficiently.

Telemedicine and Remote Consultations: Speech recognition technology can enable efficient and accurate transcription of spoken conversations between healthcare professionals and patients during telemedicine consultations. This allows for real-time documentation of patient encounters and facilitates remote collaboration and decision-making.

Clinical Documentation: Speech recognition can be used to convert spoken dictation into text, enabling healthcare professionals to create clinical documentation more efficiently. This reduces the time spent on manual data entry and allows for more accurate and timely documentation of patient encounters.

Voice-Enabled Virtual Assistants: Voice recognition technology can power virtual assistants that can respond to voice commands and provide information or perform tasks related to healthcare. These virtual assistants can assist healthcare professionals in accessing patient information, retrieving medical knowledge, and managing administrative tasks.

Natural Language Processing (NLP): NLP techniques can be combined with speech recognition to extract and analyze information from spoken conversations or transcribed text. NLP algorithms can identify relevant concepts, relationships, and sentiments, enabling advanced analysis of patient data and facilitating decision support.

Patient Monitoring and Diagnostics: Image recognition can be used in wearable devices or remote monitoring systems to analyze images or videos captured by patients. For example, image recognition algorithms can analyze skin lesions or wounds to monitor healing progress or detect changes that may require medical attention.

Quality Assurance and Compliance: Image and speech recognition technologies can be used for quality assurance and compliance purposes. For example, image recognition algorithms can analyze medical images to ensure adherence to imaging protocols or identify potential errors. Speech recognition can be used to monitor and analyze conversations for compliance with regulatory requirements or quality standards.

Accessibility and Assistive Technologies: Speech recognition can be used to develop assistive technologies for individuals with disabilities or limited mobility. For example, speech recognition can enable hands-free control of medical devices or systems, allowing individuals to interact with HIS more easily.

Implementing image and speech recognition technologies in HIS requires careful consideration of data privacy, security, and regulatory compliance. It is important to ensure the accuracy, reliability, and ethical use of these technologies to maintain patient confidentiality and trust. Collaboration between healthcare professionals, data scientists, and technology providers is crucial to develop and deploy image and speech recognition solutions that meet the specific needs of healthcare settings.

3.9 CASE STUDIES IN HEALTH INFORMATION SYSTEMS (HIS) OPTIMIZATION

Case Study 1: Electronic Health Record (EHR) Optimization

A large hospital system was experiencing challenges with its EHR system, including slow performance, inefficient workflows, and difficulty in accessing and sharing patient information across departments.

Solution: The hospital system conducted a comprehensive assessment of its EHR system and identified areas for optimization.

The hospital redesigned its workflows to streamline processes and eliminate unnecessary steps. They involved clinicians and staff in the redesign process to ensure the changes aligned with their needs. The EHR system was customized and configured to match the specific needs of different departments and specialties. This included creating templates, order sets, and clinical decision support tools tailored to each area. The hospital provided extensive training and education to clinicians and staff on how to effectively use the EHR system. This included both initial training for new users and ongoing training to keep everyone updated on system enhancements and best practices. The hospital worked on improving interoperability between the EHR system and other systems, such as laboratory and imaging systems. This allowed for seamless exchange of data and reduced the need for manual data entry. The hospital optimized the performance of the EHR system by upgrading hardware, improving network infrastructure, and implementing system enhancements recommended by the EHR vendor. The EHR optimization efforts resulted in significant improvements in efficiency and patient care. Clinicians reported faster access to patient information, reduced documentation time, and improved communication across departments. The hospital also saw a decrease in medication errors and improved patient outcomes.

Case Study 2: Data Analytics and Reporting Optimization

A healthcare organization had a wealth of data in its health information system but struggled to effectively analyze and report on it. They faced challenges in generating timely and accurate reports for decision-making and quality improvement initiatives.

Solution: The organization implemented a data analytics and reporting optimization project to address their challenges. The organization established a data governance framework to ensure data quality, standardization, and integrity. They implemented data validation processes and conducted regular data quality assessments to identify and rectify data errors. The organization developed a centralized data warehouse that integrated data from various sources, including EHRs, billing systems, and external databases. This allowed for a comprehensive view of patient and operational data. The organization implemented advanced analytics tools, such as data mining, predictive modeling, and visualization tools, to analyze and interpret the data. This enabled them to identify trends, patterns, and insights for decision-making and quality improvement initiatives. The organization automated the generation of routine reports, such as quality metrics, financial performance, and patient outcomes. This reduced manual effort and ensured timely and accurate reporting. The organization engaged stakeholders, including clinicians, administrators, and quality improvement teams, in the design and development of analytics solutions. This ensured that the reports and insights generated were relevant and actionable.

The data analytics and reporting optimization efforts led to improved decision-making and quality improvement initiatives. The organization gained better insights into patient populations, identified areas for cost savings, and implemented targeted interventions to improve patient outcomes. The automated reporting reduced the time spent on manual report generation, allowing staff to focus on data analysis and interpretation. These case studies highlight the importance of optimizing health information systems to improve efficiency, patient care, and decision-making in healthcare organizations. By identifying areas for improvement and implementing targeted strategies, organizations can leverage their HIS to its full potential.

3.10 FUTURE TRENDS AND INNOVATIONS

The field of health information systems (HIS) optimization is constantly evolving, driven by advancements in technology and the need for more efficient and effective healthcare delivery. Here are some future trends and innovations that are expected to shape HIS optimization:

Artificial Intelligence (AI) and Machine Learning: AI and machine learning technologies have the potential to revolutionize HIS optimization. These technologies can automate routine tasks, analyze large volumes of data for insights, and improve decision-making. AI-powered chatbots and virtual assistants can enhance patient engagement and support clinical workflows.

Interoperability and Health Information Exchange: Achieving seamless interoperability and health information exchange between different systems and healthcare organizations is a key focus for HIS optimization. Standards like Fast Healthcare Interoperability Resources (FHIR) are being adopted to facilitate data sharing and integration, enabling a comprehensive view of patient information across care settings.

Internet of Medical Things (IoMT): The IoMT refers to the network of medical devices and wearables that collect and transmit patient data. HIS optimization will involve integrating and analyzing data from these devices to provide real-time monitoring, personalized care, and early detection of health issues.

Blockchain Technology: Blockchain has the potential to enhance data security, privacy, and integrity in HIS. It can enable secure sharing and access to patient data, facilitate consent management, and support data provenance and auditability.

Predictive Analytics and Precision Medicine: The use of predictive analytics and precision medicine approaches will continue to grow in HIS optimization. Predictive models can help identify high-risk patients, optimize treatment plans, and improve population health management.

Telehealth and Remote Patient Monitoring: Telehealth and remote patient monitoring technologies are becoming increasingly important, especially in light of the COVID-19 pandemic. HIS optimization will involve integrating these technologies into existing systems to enable virtual consultations, remote monitoring, and telemedicine services.

Data Visualization and Dashboards: Effective data visualization techniques and interactive dashboards will play a crucial role in HIS optimization. These tools enable healthcare professionals to quickly understand and interpret complex data, facilitating data-driven decision-making.

Patient Engagement and Personal Health Records: Empowering patients to actively participate in their healthcare is a growing trend. HIS optimization will focus on providing patients with access to their health records, personalized health information, and tools for self-management.

Cybersecurity and Data Privacy: As healthcare systems become more digitized, cybersecurity and data privacy will remain critical concerns. HIS optimization will involve implementing robust security measures, encryption techniques, and privacy frameworks to protect patient data from breaches and unauthorized access.

Continuous Improvement and User-Centered Design: HIS optimization will embrace a culture of continuous improvement and user-centered design. Involving end-users, such as healthcare professionals and patients, in the design and development process will ensure that HIS solutions meet their needs and are intuitive to use.

These future trends and innovations in HIS optimization hold great potential to transform healthcare delivery, improve patient outcomes, and enhance the overall efficiency and effectiveness of healthcare systems. Embracing these advancements will require collaboration between healthcare organizations, technology providers, policymakers, and regulatory bodies.

3.11 CONCLUSION AND FUTURE DIRECTIONS

In conclusion, health information systems (HIS) optimization is a critical area of focus for healthcare organizations to improve efficiency, enhance patient care, and enable data-driven decision-making. The advancements in technology and the evolving healthcare landscape present exciting opportunities for future directions in HIS optimization.

The future of HIS optimization will be shaped by trends such as artificial intelligence (AI) and machine learning, interoperability and health information exchange, Internet of Medical Things (IoMT), blockchain technology, predictive analytics and precision medicine, telehealth and remote patient monitoring, data visualization and dashboards, patient engagement and personal health records, cybersecurity and data privacy, and continuous improvement and user-centered design.

REFERENCES

1. Pokvić, L., Spahić, L., Badnjević, A., 2020. Implementation of Industry 4.0 in transformation of medical device maintenance systems, pp. 512–532. https://doi.org/10.4018/978-1-7998-2725-2.ch023
2. Xu, Z., Wang, X., Zeng, S., Ren, X., Yan, Y., Gong, Z., 2021. Applying artificial intelligence for cancer immunotherapy. *Acta Pharmaceutica Sinica. B*, 11, pp. 3393–3405. https://doi.org/10.1016/j.apsb.2021.02.007
3. Ouadfeul, S., Jawak, S., Shirzadi, A., Idrees, M., 2023. Editorial: Artificial intelligence and machine learning in Earth science. *Frontiers in Earth Science*, 10, p. 1090016. https://doi.org/10.3389/feart.2022.1090016
4. Bennett, C., Hauser, K., 2013. Artificial intelligence framework for simulating clinical decision-making: A Markov decision process approach. *Artificial Intelligence in Medicine*, 57(1), pp. 9–19. https://doi.org/10.1016/j.artmed.2012.12.003
5. Yoldemir, T., 2020. Artificial intelligence and women's health. *Climacteric*, 23, pp. 1–2. https://doi.org/10.1080/13697137.2019.1682804
6. Walker, L., Abuzour, A., Bollegala, D., Clegg, A., Gabbay, M., Griffiths, A., Kullu, C., Leeming, G., Mair, F., Maskell, S., Relton, S., Ruddle, R., Shantsila, E., Sperrin, M., Staa, T., Woodall, A., Buchan, I., 2022. The DynAIRx project protocol: Artificial Intelligence for dynamic prescribing optimisation and care integration in multimorbidity. *Journal of Multimorbidity and Comorbidity*, 12, p. 12. https://doi.org/10.1177/26335565221145493
7. Berdutin, V., Abaeva, O., Romanova, T., Romanov, S., 2023. Achievements and prospects for the application of artificial intelligence technologies in medicine: An overview. Part 1. *Sociology of Medicine*, p. 10. https://doi.org/10.17816/socm106054
8. Khan, N., Perera, C., Dow, E., Chen, K., Mahajan, V., Mruthyunjaya, P., Do, D., Leng, T., Myung, D., 2022. Predicting systemic health features from retinal fundus images using transfer-learning-based artificial intelligence models. *Diagnostics*, 12, p. 10. https://doi.org/10.3390/diagnostics12071714

9. Jiang, F., Jiang, Y., Zhi, H., Dong, Y., Li, H., Ma, S., Wang, Y., Dong, Q., Shen, H., Wang, Y., 2017. Artificial intelligence in healthcare: Past, present and future. *Stroke and Vascular Neurology*, 2, pp. 230–243. https://doi.org/10.1136/svn-2017-000101
10. 2019. Indexing and optimization techniques in biomedical industry. *International Journal of Engineering and Advanced Technology*, 12, p. 11. https://doi.org/10.35940/ijeat.f9524.088619
11. Alattab, A., Ghaleb, M., Olayah, F., AlMurtadha, Y., Hamdi, M., Yahya, A., Irshad, R., 2022. A smart diseases diagnosis and classification strategy of electronic healthcare application using novel hybrid artificial intelligence approaches. *Journal of Nanoelectronics and Optoelectronics* . https://doi.org/10.1166/jno.2022.3355
12. Giammarile, F., Slart, R., Costa, P., 2019. Two new thematic series—Spotlight on artificial intelligence and a specific platform for technologist. *European Journal of Hybrid Imaging*, 3. https://doi.org/10.1186/s41824-019-0067-1
13. Maden, W., Lomas, D., Hekkert, P., 2023. A framework for designing AI systems that support community wellbeing. *Frontiers in Psychology*, 13, p. 11. https://doi.org/10.3389/fpsyg.2022.1011883
14. Vandenberg, O., Durand, G., Hallin, M., Diefenbach, A., Gant, V., Murray, P., Kozlakidis, Z., Belkum, A., 2020. Consolidation of clinical microbiology laboratories and introduction of transformative technologies. *Clinical Microbiology Reviews*, 33, p. 10. https://doi.org/10.1128/CMR.00057-19.
15. Nakayama, L., Kras, A., Ribeiro, L., Malerbi, F., Mendonça, L., Celi, L., Regatieri, C., Waheed, N., 2022. Global disparity bias in ophthalmology artificial intelligence applications. *BMJ Health Care Informatics*, 29, p. 10. https://doi.org/10.1136/bmjhci-2021-100470.
16. Díaz, Z., Muka, T., Franco, O., 2019. Personalized solutions for menopause through artificial intelligence: Are we there yet?. *Maturitas*, 9, p. 8. https://doi.org/10.1016/j.maturitas.2019.07.006.
17. Lake, J., 2019. Evaluating and using medical evidence in integrative mental health care: Literature review, evidence tables, algorithms, and the promise of artificial intelligence. *An Integrative Paradigm for Mental Health Care*, 1, p. 6.
18. Chang, Y., Tsai, S., Wu, Y., Yang, A., 2020. Development of an AI-based web diagnostic system for phenotyping psychiatric disorders. *Frontiers in Psychiatry*, 11, p. 7.
19. Zhang, Z., Citardi, D., Wang, D., Genc, Y., Shan, J., Fan, X., 2021. Patients' perceptions of using artificial intelligence (AI)-based technology to comprehend radiology imaging data. *Health Informatics Journal*, 27. https://doi.org/10.1177/14604582211011215
20. Gaggioli, A., 2017. Artificial intelligence: The future of cybertherapy?. *Cyberpsychology, Behavior and Social Networking*, 20(6), pp. 402–403. https://doi.org/10.1089/cyber.2017.29075.csi
21. Chen, C., Li, P., Jian, X., Wu, L., Yang, J., 2022. Optimal edge nodes deployment with multi association for smart health. *IEEE Transactions on Molecular, Biological and Multi-Scale Communications*, 8, pp. 1–8. https://doi.org/10.1109/tmbmc.2021.3118951.

4 Agricultural Information Systems Emphasizing Agro Robots Towards Digital and Sustainable Agriculture
A Scientific Review

P. K. Paul, Nilanjan Das, Rajibul Hossain, Mustafa Kayyali, and Ricardo Saavedra

4.1 INTRODUCTION

Technologies are significantly changing organizations and business houses. Both the core technologies as well as allied technologies are important for any kind of performance of the activities and that ultimately helps in services development and making production easier. Information Technology, like other sectors, is also advancing and boosting the Agricultural sector and this has significantly increased in the recent past. Various IT components are dedicated to proper and effective agricultural development. Information Technology and Computing, i.e. Agricultural Informatics helps in various agro jobs such as crops, plants management, increasing speed management and systems of the agricultural affairs. Some of the other activities such as in disease detection as well as sustainable agricultural development are also significantly increasing the agro-space including agricultural products, etc. [2], [3], [38]. Artificial Intelligence is also enhancing the agricultural systems in different areas, and it helps in the productivity and quality enhancement of agro products and ultimately helps in developing crop marketing-related matters. The AI has tremendous potential in agricultural systems and this can be achieve using drone-based image processing, proper and effective precision farming, etc. The bodies and organizations who are associated with agriculture can use Artificial Intelligence and Robotics for wider and healthy agricultural systems [47]. Today we are moving towards smart towns, smart villages and into the digital society and all these are supported by different tools and technologies. Today, agricultural sciences are highly accustomed to scientific and technologically allied systems. Artificial Intelligence is dedicated to developing intelligent systems and such empowered machines are dedicated to agricultural systems for pre- and post-production-related activities. All such mechanisms are called intelligent agents. AI is dependent on various types of machines thus it is also known as Machine Intelligence and it can also do the problem-solving of agriculture. Computational statistics are useful in post-production activities of agricultural systems. Mathematical optimization with proper and effective theory and applications is worthy in developing sustainable agricultural systems. Robotics is not only a tool or technology now, rather it has become an interdisciplinary field of study relying on intelligent systems more than traditional components of IT and depends on many of the latest technologies viz. Cloud and Fog Computing, Artificial Intelligence, Big Data and Analytics, Robotics, HCI and Usability Engineering, and so on [11], [39].

4.2 OBJECTIVE OF THE WORK

The work entitled "Agricultural Information Systems emphasizing Agro Robots towards Digital and Sustainable Agriculture – *A Scientific Review*" is aimed at following core aspects.

- To gather basics of Agricultural Information Systems with its fundamental features, and functions.
- To learn about the allied concern of Agricultural Information Systems (AIS) which include Agricultural Information System Management, Agricultural Web Portal Management, and Agricultural Information Management Standards.
- To learn about the basic and emerging applications of AI and Robotics in Agricultural Systems.
- To learn about Agricultural Robots and their types in the contemporary context of advancing agricultural systems.
- To find out the future potentialities and possibilities of Agro AI and allied concerns in agricultural systems.

4.3 METHODS

As far as the work titled "Agricultural Information Systems Emphasizing Agro Robots towards Digital and Sustainable Agriculture – *A Scientific Review*" is concerned this is completely a theoretical work and lies on existing sources of literature and resources. Various books, journals, periodicals, and magazines related to Agricultural Information Systems including Agricultural Information Networks, Agricultural Web Portal Management, standards related to the Agricultural Information Management are studied, analyzed and reported in this work.

4.4 RELATED EXISTING WORKS

As the work is based on a review of the literature and existing resources various consulted works are here reported and among them, a few important ones are as follows.

Chaganti et al. (2022) [6]. In their work, they focus on the innovative integration of blockchain and Internet of Things (IoT) technologies to enhance security monitoring in the realm of smart agriculture. The paper acknowledges the rising importance of advanced technological solutions in agriculture, particularly in addressing the growing concerns over security and data integrity in the sector. At the heart of this exploration is the concept of 'smart agriculture' – a forward-thinking approach that incorporates cutting-edge technologies to increase the efficiency and sustainability of agricultural practices. The authors examine how the convergence of blockchain and IoT can create a robust framework for security monitoring. Blockchain's inherent features of decentralization, immutability, and transparency provide a secure and reliable foundation for managing agricultural data. IoT devices provide the ability to monitor and collect data in real-time, which is crucial for contemporary agricultural methods. The paper explores the possibility of cooperation among various technologies. IoT devices have the capability to gather extensive quantities of data from the agricultural sector, encompassing measurements such as soil moisture levels, weather conditions, and crop development patterns. This data, when recorded on a blockchain, becomes tamper-proof and easily verifiable, enhancing the trustworthiness of the information and the decisions based on it. In summary, this paper presents a comprehensive look at how the integration of blockchain and IoT can revolutionize security monitoring in smart agriculture. Offering a detailed analysis of the benefits and challenges, it contributes to the ongoing discussion on the future of agricultural practices and the role advanced technologies can play in making agriculture more secure, efficient, and sustainable. Ferrández-Pastor et al. (2022) [13]. The paper represents a substantial advancement in the modernization of agricultural methods and the guarantee of the honesty and openness of the supply chain. The authors commence by addressing the growing need for traceability in the field of agriculture. Traceability pertains to the capacity to systematically monitor and authenticate the past events, current whereabouts, and projected path of a product. In the context of agriculture, and particularly in sensitive sectors like industrial hemp, ensuring traceability is crucial for complying with legal standards, ensuring product quality, and maintaining consumer trust. The paper then introduces the concept and components of the proposed traceability model. The utilization of IoT devices is essential in this context, as they can be strategically placed across the agricultural and production cycle to gather up-to-the-minute information on diverse factors including crop development conditions, harvesting schedules, and processing phases. This data is extremely valuable for establishing a thorough and transparent documentation of the product's whole trajectory from the farm to the consumer. The paper reflects on the broader implications of this model for the agricultural sector and beyond. This includes discussions on the potential scalability of the model to other agricultural products, the economic and social benefits of improved traceability, and the challenges and considerations that need to be addressed for wider adoption. Overall, the paper offers a promising glimpse into the future of agriculture, where advanced technologies like IoT and Blockchain work hand in hand to ensure the integrity, safety, and sustainability of our food supply.

Khan et al. (2022) [20] also investigate the deployment of blockchain technology in agricultural supply chains, particularly focusing on the period of the COVID-19 pandemic. The paper emphasizes not just the utility and efficiency of blockchain in this critical sector but also its role in promoting cleaner, more sustainable practices. The authors begin by outlining the unprecedented challenges that the COVID-19 pandemic posed to agricultural supply chains worldwide. These challenges may include disruptions due to lockdowns, labor shortages, logistical constraints, and heightened demand for transparency and safety from consumers. The pandemic underscored the vulnerability of traditional supply chains and the urgent need for more resilient, flexible, and transparent systems. The paper then shifts focus to blockchain technology, describing its fundamental features and how it can address the challenges posed by the pandemic. The authors discuss how blockchain's decentralized nature allows for more resilient supply chains that are less dependent on central authorities and more capable of adapting to disruptions. They could also explore how the technology's ability to create a tamper-proof and transparent record of transactions can enhance traceability and accountability, crucial for ensuring the safety and quality of food products during a public health crisis. A significant portion of the paper is dedicated to the specific benefits and cleaner solutions that blockchain offers for agricultural supply chains during the pandemic. This includes reducing waste by more efficiently matching supply with demand, promoting more sustainable practices by providing consumers with clear information about the origin and journey of their food and reducing the need for physical paperwork and interactions, which can help limit the spread of the virus. The paper reflects on the role of technology in addressing global crises and promoting sustainability. The authors argue that while blockchain is not a panacea, it offers valuable tools that, when combined with other measures, can significantly enhance the resilience, efficiency, and sustainability of agricultural supply chains. Overall, this paper provides a timely and insightful exploration of how cutting-edge technology can be harnessed to address some of the most pressing challenges facing the world today, offering a valuable resource for researchers, policymakers, and industry practitioners alike.

Mangla et al. (2022) [29] highlighted in their paper a deep exploration of how blockchain technology might be leveraged to construct sustainable supply chains, with a special focus on the tea sector. This study proposes a conceptual framework meant to guide the deployment of blockchain in promoting sustainability and also identifies potential challenges that limit its adoption. The authors begin by discussing the growing importance of sustainability in supply chain management. They elaborate on the various challenges that industries face in trying to achieve sustainability goals, such as lack of transparency, inefficiencies, and the difficulty in verifying the ethical and environmentally friendly practices of suppliers. The paper then offers blockchain technology as a viable answer to these difficulties. The authors explore how the intrinsic qualities of blockchain, such as immutability, transparency, and security, make it well-suited to developing more sustainable and dependable supply chains. They might also provide an overview of how blockchain has been utilized in other sectors and the potential benefits it can provide. The crux of the study is the description of a conceptual framework for creating a blockchain-based a sustainable supply chain. This framework outlines the key components and steps involved in integrating blockchain into supply chain operations, from the initial assessment of needs and challenges to the development and deployment of blockchain solutions. The authors discuss how this framework can be applied specifically to the tea industry, considering the unique characteristics and requirements of this sector. The paper reflects on the potential of blockchain to transform supply chains and promote sustainability. The authors underline the necessity of addressing the identified challenges and offer advice for enterprises, legislators, and researchers on how to promote the effective deployment of blockchain-based supply chain solutions. Overall, this paper presents a comprehensive and practical guide to utilizing the power of blockchain for sustainable supply chain management, delivering significant insights and a clear methodology that can be applied to the tea sector and beyond. **Mavilia and Pisani** (2022) [30] in their work provide a focused examination of how blockchain technology can be applied within the agricultural sector of South Africa. This paper explores the specific context, challenges, and opportunities of adopting blockchain in South African agriculture, offering insights that may apply to similar regions and sectors. The authors begin by outlining the significance of the agricultural sector in South Africa, discussing its role in the economy, the livelihoods it supports, and the unique challenges it faces. These challenges include issues related to supply chain inefficiencies, lack of access to markets, land ownership disputes, and the need for sustainable practices. The paper then presents blockchain technology, detailing its basic concepts and how it may address some of the difficulties encountered by the agricultural sector. The authors highlight the potential of blockchain to improve traceability and transparency, cut transaction costs, and enhance the security of data and transactions. The main focus of the study is an analysis of how blockchain technology can be specifically implemented in the South African agricultural environment. This analysis considers the current state of technology infrastructure in the country, the readiness of the agricultural sector to adopt new technologies, and the specific needs and priorities of South African farmers and agricultural businesses. The paper reflects on the potential impact of blockchain on South African agriculture, considering both the opportunities and the challenges. The authors present ideas for governments, industry leaders, and researchers on how to assist the successful use of blockchain technology in the sector. Overall, this paper offers an invaluable boost to recognizing how blockchain technology can be applied in a specific regional and sectoral context, offering ideas and suggestions that can help guide the establishment of more effective, transparent, and environmentally friendly agricultural systems in South Africa and similar contexts.

Nayal et al. (2023) [35] work and highlights the in-depth examination of the factors and conditions that facilitate the implementation of blockchain technology in creating sustainable agricultural supply chains. This scientific study looks into the conditions and influential variables essential to properly integrate blockchain technology into agriculture, focusing on developing a supply chain that is not only efficient but also sustainable. The authors begin by contextualizing the importance of sustainability in the agricultural supply chain. They examine the growing global demand for food and the associated need for agricultural practices that are environmentally benign, commercially feasible, and socially responsible. The report addresses the problems in doing this, such as inefficiencies in the supply chain, waste, fraud, and the lack of transparency and traceability. Following this, the study proposes blockchain technology as a viable answer to these difficulties. The authors describe how blockchain's qualities – including decentralization, immutability, and transparency – make it well-suited to increasing the sustainability of agricultural supply chains. They might present an overview of how blockchain has been used in other sectors and the potential benefits it can bring to agriculture. The centrepiece of the research is a detailed investigation of the antecedents for integrating blockchain in the agricultural supply chain. These antecedents are the required circumstances or precursors that must be in place for blockchain technology to be successfully deployed. The authors split them into many sorts, such as technological elements (e.g., infrastructure and knowledge), organizational factors (e.g., leadership and strategy), and environmental factors (e.g., regulatory backing and market demand). The article comments on the potential of blockchain to alter agricultural supply chains, making them more sustainable and resilient. The authors underline the necessity of recognizing and resolving the antecedents for blockchain implementation, proposing a roadmap for future study and practice in this field. Overall, this research contributes to the scholarly literature on blockchain and agriculture by giving a nuanced and extensive analysis of the variables necessary for successful technology integration. It gives excellent insights for anyone interested in the nexus of technology, agriculture, and sustainability.

Araújo et al. (2021) [1] describe the future trends and present application domains, challenges, sustainability of Agriculture 4.0. Agriculture 4.0 is being implemented using technologies including sensors, robotics, Internet of Things, Cloud Computing, Data Analytics, and Decision Support System. Agriculture 4.0 has various applications in supply chain system of production and distribution such as production, transportation, storage, processing, distribution, and consumption. Domains of Agriculture 4.0 implementations are monitoring, control, prediction, and logistics. There are several challenges to implementing Agriculture 4.0, such as cost, maintaining quality, availability, privacy, data security, scalability, hardware and software compatibility, integrity, complexity, cyber security, and other issues as per the technology, application and system. **Barakabitze et al.** (2017) [3] emphasize the action research to support communities in farming using the aid of Information Communication Technology. ICT is significant in the improvement of the farming process. Therefore, the main aim is to raise awareness about ICT implementation in farming. It has been observed that some of the ICT applications are not used by the farmers. Different techniques of participatory action research are applied to conducting the research and development in the utilization of ICT solutions in farming. **Bechar and Vigneault** (2016) [4] show the robotics application in agricultural fields. Robots are automated machines equipped with sensor devices, processors, and software. The aim of this study is to find out how the robots perform and what is the progress of robots in the field of agriculture. The implantation cost is high to utilize robotic systems in farming. Nowadays, technology is more advanced in nature but there are barriers to implementing technology in the farming field. Robots in the field of farming collect information through the sensor devices and send information for further analysis and research. Robots are also used in farming processes such as tree plantation, harvesting, monitoring the growth of crops, and watering the plants.

Behera et al. (2015) [5] point out improvements in the marketing sectors of the agricultural field by using ICT. India has a huge prospect in the agricultural retail market because it is the second-largest country in the agricultural field of production. It shows what is the significant role of ICT in the retail business of agriculture. **Channe et al.** (2015) [7] approach a model that is built using advanced technologies such as the Internet of Things (IoT), Cloud Computing, AI tools for Big Data analysis, and Mobile Computing. Other stakeholders in this model are farmers, Agro-vendors, and Agro-agencies. Sensors in IoT are used to capture the real-time data from agricultural fields like soil condition for farming and environment status. Sensors capture data and send it to the Agro cloud for further analysis of the data using Big Data AI tools. To prepare dataset, data has been collected from harvesting and farming fields.

Chebrolu et al. (2017) [8]. Agricultural Robots are used to collect data from the farming fields. Nowadays automated sensor-enabled robots are very popular in the agricultural sector for harvesting and also collection of data for further research and development in this sector. The robots hold RGB-D sensors and also multichannel cameras to monitor and capture valuable information from the fields. For the improvement of farming, farming-related valuable data is precious. The main target is analysis of data to localize mapping and classify sugar beet fields. Agricultural Robotics are involved in farming and in harvesting fields to provide services of accurate harvesting. Robots are very useful for different purposes of farming. **Cheein et al.** (2013) [9]. Autonomous or unmanned robots are also involved in the field of agriculture to monitor the weed effect in crops, leveling of terrain, irrigation, and dispersal of agrochemicals. Therefore, robots do all the tasks that farmers can do, very efficiently. Autonomous robots are classified in different domains such as guidance, detection, action, and mapping. Guidance focuses on the navigation path in which robots are moved within the boundaries of the agricultural field. In detection, robots are used to collect biological information from fields. In action, robots are actively used in harvesting. In mapping, maps of the surrounding agricultural fields are developed by robots. Therefore, robots provide services related to farming which help the farmer for the betterment of farming. **Da Silveira et al.** (2021) [10] show how Agriculture 4.0 is being implemented and established without any barriers but also that many barriers are faced in implementing Agriculture 4.0. Agriculture 4.0 has lots of benefits to improve the agriculture in this era. There are many advantages, disadvantages, and barriers of implementation of Agriculture 4.0. The barriers are mainly from economic, social, and environmental aspects. Agricultural Information give awareness about the production. It gives information about the land, livestock, labour, capital and management. The information is shared between the farmers. Farmers get all the benefits regarding farming from the agricultural information system. Agricultural information improves the farming and productivity.

4.5 AGRICULTURAL INFORMATION SYSTEM MANAGEMENT

The Agricultural Information Management System (AIMS) is a digitalized system that delivers services and provides several farming-related facilities to the farmer. Farmers can get all the benefits regarding farming from a wide range of Agricultural Information Management Systems. The Department of Agriculture gives various services to the farmer through a web portal [12], [19], [42]. It gives a platform where planners, policymakers, and other stakeholders can get access to agricultural-related information. The Agricultural Information Management System gives paperless information services to the farmer including advisory, reporting of crop diseases, certificate, scheme-based, grievance redressal, and data collection from department offices. Farmers obtain numerous facilities concerning the observation of farming progress and also have the ability to update the information through a dashboard. Facilities provided to the farmer using AIMS are as follows:

- Insurance of crops – farmers can get insurance services.
- Revenue Land Information System (ReLIS) is the portal to verify land records.
- Collaborated with PMKISAN.
- AIMS gives support for crop damage due to natural disaster.
- Revenue provided to the owner for cultivable lands.
- Mobile apps available for the farmer with two regional languages.

The features of the AIMS include a personalized dashboard for farmers, declaration of land cultivation details, online application submission, e-bill submission, direct benefit transfer to farmers, digitally signed insurance policy certificate, onetime self-registration by farmer, tracking of application status, alert provided to the farmer on inspection date, approval or rejection of application, push notification given to mobile phones, automated system for calculating premium amount, insurance benefits, and compensation [4], [9], [49]. These facilities are available to the farmer from the mobile application or website portals of AIMS. Advantages of the services given by the Agricultural Information Management System include the farmer getting all the government services online without availing services from offices, a decrease in application processing time, benefits transferred to the farmer directly to the account, an online account opening facility, and a centralized DBT system for allocation of fund to lower offices. These are the advantages of using the Agricultural Information Management System in agricultural sectors to foster and improve agriculture which helps to develop the economy of the country.

4.6 AGRICULTURAL INFORMATION MANAGEMENT STANDARDS

The Food and Agriculture Organization of the United Nations (FAO) is the organization that provides world-wide expert community in food, agricultural information and data management services through a website portal. information management standards, techniques, and tools shared and accessed among all the information workers all over the world and develop universal community services [14], [16], [53]. AIMS provides a platform to share knowledge and information regarding agricultural sectors between groups of people all over the world. The FAO has a website portal where people can join freely as members in the community to share the best projects, learning facilities, standards, innovative unique ideas, questions and outcomes, and events. AIMS usage is growing and approximately 4,500 experts are involved in data management and agriculture information in whole world. This online forum gives a common platform where participants engaged in communication, share their views, thoughts and share information globally. The main aims of FAO are food safety and security, improve nutrition, enhance production growth in agriculture, and extend standards of living in rural areas [9], [18], [41]. There are three categories under AIMS including vocabulary, tools and services. AGROVOC, AgMES, and Linked Open Data are the vocabularies under AIMS. AgriDrupal, AgriOcean, VocBench, and WebAGRIS are the tools in AIMS. AGRIS, AgriFeeds, and E-LIS are the services under AIMS. The variety of Vocabulary, Tools and Services of Agriculture Information Management Standards are as follows.

- AGROVAC is a multiple regional languages vocabulary which has 20 languages and 40,000 concepts and fulfils subject including agriculture, forestries, and fisheries. It is managed by FAO and it is also linked open data set of agriculture. It is freely used in multiple languages across the globe.
- Linked Open Data is the vocabulary application that is used to give assistance to providers of bibliographic data to utilize proper encoding methodology as per their requirement in metadata exchange and formatting of bibliographic data.
- AgriDrupals is the tool of AIMS. It is applied for managing information of agriculture and community services. Agriculture Community build up with the peoples who are from various locations in world as efficient groups of people. Community are efficient in agriculture information management.
- AgriOceanDspace is developed by the Food and Agriculture Organization of the United Nation (FAO) and UNESCO-IOC/IODE to give modified Dspace version. AgriOcean is developed for metadata, and other control vocabularies with regards oceanography, marine science food, fisheries, forestry's, natural resources, agriculture, and many others.
- VocBench is the web portal of vocabularies management tools invented by FAO. It gives multiple languages facilities. The objective of the VocBench tools is to convert thesauri, authority lists, and glossaries into SKOS or RDF concepts of schemes to utilize in a linked data environment. It also helps to control workflow and editorial process referred by evolution of vocabulary included rights, validation and versioning.
- WebAGRIS is a multi-language web-based application that is used to distribute, data input, and processing of bibliographic agriculture information [17], [40], [52].
- AgriFeeds is an AIMS service which is implemented for helping users to extract news and event related agriculture information from various agriculture sources. Customized feeds are developed after collecting the information from sources.

- AGRISis also a service of AIMS which is a large public data storage consisting of 3 million bibliographical records related to agriculture science and technology. Data in the database is created by over 100 institutions from 65 countries and which is managed by FAO.
- E-LIS is the global Electronics Library and Information Sciences service and it is free and open access. It is controlled and managed by the 73 libraries international team from 47 countries.

4.7 AGRICULTURAL WEB PORTAL MANAGEMENT

Agriculture Web Portal referred websites portals which are used to deliver services related to agriculture. Farmers use web portals to access services provided by agricultural departments or organizations. The main objective of agriculture web portal management is to manage all the agriculture information related work and services through websites portal. Data collected from agricultural fields are stored in repository for further utilization. Data accessed and managed using website portals. In agricultural fields, data collected and stored in online data store from various sources like catalogues of crops, information of crops diseases, production of crops, research and experiment data and others.

Some of the web portals are used to maintain catalogues of plants. Like SINGER is the plants catalogues management website portal under CGIAR for research and development. This web portal maintains plants genetic information. EURISCO is the online catalogue web portal of European plants genetic information as biodiversity data [23], [24], [51]. The Germplasm Resources Information Network (GRIN) is a web portal maintained by the United States of America Department of Agricultural Research Service to manage data storage of genetic diversity information of significant plants of agriculture (Arnaud et al. 2010). Therefore, all the web portals related to agriculture information are maintained and managed globally by various organizations and institutions for further utilization and research purposes.

4.8 AI AND AGRO SYSTEMS

Agricultural Informatics is supported by different Information Technology components and out of which Artificial Intelligence is important, which is supported by various tools, technologies and systems. Machines is very important in developing intelligent and effective intelligent products. AI is helpful in human speech, and for advancing strategic systems and automated operating cars, which are strongly supported by knowledge reasoning, ML and DL, including computer vision, NLP, etc. Here, Figure 4.1 (designed by authors) depicts the basic approaches of AI.

Artificial Intelligence is closely dependent on robotics and vice versa, further Data Mining is also connected with AI. Artificial Intelligence is dependent on exploratory data analysis and this includes unsupervised learning, predictive analytics, etc. AI is interconnected with other sub-systems which are dedicated in the development of various products, services and systems. Artificial Intelligence is helpful in healthy performance building like a human, further also able to mimic the action. Artificial

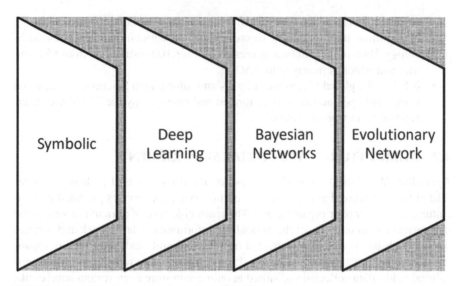

FIGURE 4.1 Basic Approaches in Artificial Intelligence.

Intelligence is thus learning and then doing the problem-solving for the proper and specific goal [20], [21], [27]. Artificial Intelligence is being used in different sectors such as healthcare and medicine, governance and administration, business industries, transportation of various kinds, education, learning and teaching. Artificial Intelligence is lies in automated machines, and widely increasing in agricultural sectors for higher capacity and proper growth. Artificial Intelligence is generally considered as several features and functions such as –

- AI relies on simulation similar to the intelligence of various devices and machines.
- In learning, reasoning, as well as in perception AI and allied systems are effective.
- Depending upon need, weak AI or strong AI is applicable.

In general, Artificial Intelligence is of two types: weak AI and strong AI. And as far as the complexity is concerned, strong AI is important, though weak Artificial Intelligence is useful in simple situations and in less AI-required devices. In more advanced Artificial Intelligence-dependent systems a robot is important, and this is treated as a more complex task, and treated as human-like. It is more complicated therefore it solves many basic problems.

4.9 AGRICULTURAL ROBOT: BASICS AND TYPES

4.9.1 Harvesting Robots

Robots that have been developed to simplify harvesting are called harvesting robots. These robots are mainly used for picking or harvesting vegetables or crops. A variety of technologies have been used to build such robots. The technologies that are used are computer vision and machine learning. Using this technology, robots can easily identify ripe crops. As a result, harvesting tasks can easily be done by such robots. These robots help meet the shortage of skilled labour required in the agricultural sector [46], [50]. This harvesting robot can easily understand the subtle differences between many crops and which are suitable for harvesting in any weather. Fruits and vegetables like tomatoes and strawberries are some of these fruits or crops that are very difficult to identify when they are ripe. They have to be picked up very gently and delicately and placed in a specific container. Harvesting robots simplify this complex process for the farmers and make them more efficient.

4.9.2 Weeding Robots

Weeding in agriculture is one of the major concerns of farmers. Weeding robots are specially made keeping the farmers in mind. These types of robots are designed in such a way that they can easily identify weeds and remove those weeds along with it. With computer vision AI and machine learning, such robots have been developed in such a way that such robots can easily differentiate between crops and weeds. When farmers do weeding, either by themselves or by labourers, the weeding affects the surrounding crop plants [38], [48]. Also, crops are damaged in the process of killing weeds by spraying pesticides physically. Sometimes even crops are damaged due to wrong pesticides or too little or too much pesticide. But these weeding robots can work day and night in any weather. The use of such weeding robots has increased crop production. The use of pesticides has decreased resulting in less pollution to the environment. But these weeding robots will definitely play a crucial role in revolutionizing the agricultural sector in the future.

4.9.3 Seeding and Planting Robots

Seed sowing is an important part of the crop production process in agriculture. The role of seeding and planting robots in agriculture is immense. This type of robot can easily sow seeds anywhere from fields to nurseries in any weather condition. These types of robots are mainly developed using technologies like AI and its vision. How a crop will produce depends on the seed. If the seeds are sown in a proper manner then the yield of the crop will be good. And these seeding and planting robots sow seeds very easily and efficiently, resulting in better crop production. The use of such robots fills the shortage of skilled workers and reduces costs a lot.

4.9.4 Spraying and Pest Control Robots

Pests are a major problem in agriculture. So farmers have been using many pesticides to control this pest for a long time. When this pesticide is being used by humans, the person is being harmed or many physical problems may occur, besides, many crops are damaged. Therefore, a type of robot for spraying and pest control has been developed for use in agriculture. The main function of such robots is aerial spraying of pesticides on crops and pesticide control. Technologies used to power these robots include machine learning, Artificial Intelligence, and various types of sensors. The main purpose of developing such Spraying and Pest Control Robots is to help farmers use pesticides properly and protect the environment. Also, this type of robot has some additional features that are cameras and sensors. As a result, such robots can provide the farmer with accurate information on both the exact number of pests and the health of the crop. As a result, farmers can make better decisions and increase their crop production. The use of spraying and pest control robots can reduce labour shortage and manual labour.

4.9.5 Sorting and Packing Robots

Sorting and packing robots the main reason for making this type of robot is to sort the products along with cutting and packing the crops. Such robots are designed in such a way, and with the use of new machine learning technology, that they can easily pick the crop colour based on crop size and crop estimation. These robots are of such good quality that they can even pick vegetables [40], [49]. Once the batch phase is over, the robot adjusts and starts packing. These robots are designed in such a way that they can easily pack the crops with great care without causing any bruises or damage. However, it is becoming possible to commercialize agriculture better for the use of such robots. Robots like these will further enhance agriculture in the future.

4.9.6 Soil Sampling Robots

Agriculture is mainly done on land. And Soil Sampling Robots have been created to know the correct information about this soil. These types of Soil Sampling Robots mainly provide information about soil health, fertility and soil structure. As a result, the farmers can easily understand what crops will be produced on which land, which fertilizers should be applied or how much fertilizer should be applied, along with how much irrigation or water should be given so that good crops can be produced easily. These types of robots are designed using technologies like sensors, cameras and Artificial Intelligence in such a way that they can provide accurate information in any situation and in bad weather. Such robots provide field information through various methods, one of which is spectrophotometer, also field component analysis in terms of fluid absorbance values. Soil Sampling Robots will play a significant role in improving modern agriculture.

4.9.7 Autonomous Greenhouses Robots

Generally, Autonomous Greenhouse Robots are developed to take care of greenhouse farming. Such robots have been able to transform greenhouse farming into an advanced agriculture thanks to their precision efficiency. These Autonomous Greenhouse Robots use technologies such as AI robotics and IoT etc. Such robots basically control the temperature, humidity, and lighting of the greenhouse and how to control them. If the greenhouses are kept under human control, no human being can monitor them 24 hours a day, resulting in operational disruptions. But Autonomous Greenhouse Robots can monitor and operate greenhouses 24 hours a day. Using these robots has reduced the initial cost of greenhouse farming. Moreover, the need for skilled workers to operate and maintain greenhouses has been a major problem in greenhouse farming, and Autonomous Greenhouse Robots have solved this problem.

4.9.8 ANIMAL HUSBANDRY ROBOTS

The contribution of Animal Husbandry Robots in modern animal husbandry is immense, animal husbandry is seeing a new direction through the use of these robots. These Animal Husbandry Robots are mainly designed to solve all the problems of animal husbandry [15], [53]. The technologies used to build these robots are robotics AI sensors, etc. Apart from these technologies, the use of data science has also been used in this robot. These robots mainly feed the animals, clean them, monitor their health and care for them. Such robots pump milk from dairy cows and monitor the quality and quantity of that milk. These robots have cameras and sensors that track the movement of animals and alert the owner when they notice any abnormality or problem in the behaviour of an animal. These robots also do everything from cleaning the poultry houses to giving them timely food vaccines. These Animal Husbandry Robots have played a significant role in commercializing animal husbandry.

4.10 SCENARIO OF AGRICULTURAL ROBOTS AND EMERGENCE

Robotics is dependent on and maintained by Artificial Intelligence as well as in many contexts by Expert Systems. Robotics is simply a programmable computer and all these are automatically performed by various tasks [22], [26], [43]. Robotics relies on integrated systems and in a machine, which could be built inside or outside. A robot may look like a human or have a different form or shape. Today robots are being used in proper and healthy agriculture systems, and this may include the activities of pre-production as well as post-production activities of agriculture. Some of the activities of agricultural systems are dedicated in proper harvesting, efficient watering, post agro-marketing, etc. As far as drones are concerned, they may also be classed as a type of 'Robotics' which is helpful in bringing the following –

- Weed and crop controlling
- Plant seeding and developing
- Assessing the environment
- Monitoring agricultural environment
- Analyzing the soil with its proper mapping, and
- Monitoring crops in soil development stages with proper analysis, etc.

Information Technologies many latest viz. Cloud and Fog Computing, Big Data and Analytics, UXD and HCI, and importantly Robotics and AI, etc. are considered as worthy and impactful. Agriculture is important for all of us and proper data is important in some of the activities like mapping weather conditions including analyzing the temperature, water usage and levels, and proper analyzing of the soil conditions, etc. The real-time data is important in modern systems of agriculture for farmers to generate greater quantities and more varieties of crop. As far as precision agriculture is concerned, Artificial Intelligence is valuable for detecting diseases of the plants as well as pests. Regarding Artificial Intelligence-based sensors, data collection (real-time) is important for various jobs viz. agricultural accuracy as well as productivity. Further, development and predicting weather patterns are possible with Robotics for further decision-making of the farmers and agricultural merchants [25], [31], [36]. Seasonal forecasting is important in some of the countries, specially in developing countries, and in this context use of Robotics is considered as important as small farms basically produce 70% of the crops in the world.

Artificial Intelligence and Robotics are highly important in finding ground-level data as well as for monitoring using drones, etc. Thus in various agricultural Robotics as well as other allied technologies viz. Computer vision, ML & DL are important for processing data which are captured from drones and applicable for farm management. As far as unmanned drones are concerned they save time and are suitable for monitoring, where some of the devices and systems like Chatbots are considered as valuable for better and effective Agricultural Informatics practice [12], [28], [32].

Agro Informatics and its practices are highly required as the worldwide population is forecast to reach about nine billion within the year 2025, as per scientific reports. Therefore many countries are emphasizing their best in starting Agro Informatics implementation including Agro Robots and allied activities [44], [45]. As a whole in Agricultural Robots, the following are growing rapidly:

- Intelligent precision seeding
- Harvesting-based agro robots
- Robots for spraying (micro and mobile)
- Robotic Automation Process
- Robot powered by LiDAR
- Intelligent drones
- AI-based precision irrigation system
- Marketing and packing based robots
- Monitoring robots, etc. [33], [34], [37].

The next generation of farming is also important and properly introduced in many countries with the initiation of disruptive technologies, specially robotics and allied technologies. Since many countries use robots in their business, organizations and industries, for the agricultural sector also Robots and allied technologies are highly impacting and growing drastically. In the USA, New Zealand, Canada, South Korea, Japan, and some other countries robotics are highly practiced and emerging rapidly.

4.11 CONCLUDING REMARKS

Artificial Intelligence, Machine Learning and DL including some of the other allied technologies are helpful in support of agriculture. Agricultural Informatics brings a lot of opportunities and at the same time it also comes with some deficiencies; all these can be waived using Artificial Intelligence, and in all these activities Robotics and AI systems are impactful. Robotic-related activities today include activities such as analyzing weather conditions or finding the presence of pests. Planning-related matters are also important in proper Agricultural Informatics practice and some of the allied activities included in harvesting Artificial Intelligence and Robotics are considered as crucial. As Agricultural Informatics is highly supported by IT systems the latest Artificial Intelligence and Robotics are highly important in further agricultural development. Agricultural Robots are effective and offer proficiency for better processes. Mobile robotics is an important activity and rising rapidly as far as the agricultural sector is concerned. Mobile robotics may also be introduced in developing countries in addition to developed countries using proper strategies and planning. Proper and improving productivity, environmental sustainability, etc. are positively possible with Agricultural Informatics practice emphasizing Agricultural Robots. Further factors related to the labour shortage and agricultural insurance are achieved with Agro Informatics.

REFERENCES

[1] Araújo, S. O., Peres, R. S., Barata, J., Lidon, F., & Ramalho, J. C. 2021. Characterising the agriculture 4.0 landscape—Emerging trends, challenges and opportunities. *Agronomy*, *11*, 667.

[2] Arnaud, E., Dias, S., Mackay, M., Cyr, P., Gardner, C., Bretting, P., ... & Louafi, S. 2010. A global portal enabling worldwide access to information on conservation and use of biodiversity for food and agriculture. In: M. Lisa & T. Klaus (eds.), *Information and Communication Technologies for Biodiversity Conservation and Agriculture* (pp. 168–180). Aix-la-Chapelle: Shaker Verlag.

[3] Barakabitze, A. A., Fue, K. G., & Sanga, C. A. 2017. The use of participatory approaches in developing ICT-based systems for disseminating agricultural knowledge and information for farmers in developing countries: The case of Tanzania. *The Electronic Journal of Information Systems in Developing Countries*, *78*, 1–23.

[4] Bechar, A., & Vigneault, C. 2016. Agricultural robots for field operations: Concepts and components. *Biosystems Engineering*, *149*, 94–111.

[5] Behera, B. S., Panda, B., Behera, R. A., Nayak, N., Behera, A. C., & Jena, S. 2015. Information communication technology promoting retail marketing in agriculture sector in India as a study. *Procedia Computer Science*, *48*, 652–659.

[6] Chaganti, R., Varadarajan, V., Gorantla, V. S., Gadekallu, T. R., & Ravi, V. 2022. Blockchain-based cloud-enabled security monitoring using internet of things in smart agriculture. *Future Internet*, *14*, 250.

[7] Channe, H., Kothari, S., & Kadam, D. 2015. Multidisciplinary model for smart agriculture using internet-of-things (IoT), sensors, cloud-computing, mobile-computing & big-data analysis. *International Journal of Computer Technology & Applications*, *6*, 374–382.

[8] Chebrolu, N., Lottes, P., Schaefer, A., Winterhalter, W., Burgard, W., & Stachniss, C. 2017. Agricultural robot dataset for plant classification, localization and mapping on sugar beet fields. *The International Journal of Robotics Research*, *36*, 1045–1052.

[9] Cheein, F. A. A., & Carelli, R. 2013. Agricultural robotics: Unmanned robotic service units in agricultural tasks. *IEEE Industrial Electronics Magazine*, *7*, 48–58.

[10] da Silveira, F., Lermen, F. H., & Amaral, F. G. 2021. An overview of agriculture 4.0 development: Systematic review of descriptions, technologies, barriers, advantages, and disadvantages. *Computers and Electronics in Agriculture, 189*, 106405.
[11] Demiryurek, K., Erdem, H., Ceyhan, V., Atasever, S., & Uysal, O. 2008. Agricultural information systems and communication networks: The case of dairy farmers in Samsun province of Turkey. *Information Research, 13*, 13–2.
[12] Edan, Y. 1995. Design of an autonomous agricultural robot. *Applied Intelligence, 5*, 41–50.
[13] Ferrández-Pastor, F. J., Mora-Pascual, J., & Díaz-Lajara, D. 2022. Agricultural traceability model based on IoT and Blockchain: Application in industrial hemp production. *Journal of Industrial Information Integration, 29*, 100381.
[14] Foglia, M. M., & Reina, G. 2006. Agricultural robot for radicchio harvesting. *Journal of field Robotics, 23*(6–7), 363–377.
[15] Gómez-Chabla, R., Real-Avilés, K., Morán, C., Grijalva, P., & Recalde, T. 2019. IoT applications in agriculture: A systematic literature review. In *2nd International Conference on ICTs in Agronomy and Environment* (pp. 68–76). Cham: Springer.
[16] Goraya, M. S., & Kaur, H. 2015. Cloud computing in agriculture. *HCTL Open International Journal of Technology Innovations and Research (IJTIR), 16*, 2321–1814.
[17] Javaid, M., Haleem, A., Singh, R. P., & Suman, R. 2022. Enhancing smart farming through the applications of Agriculture 4.0 technologies. *International Journal of Intelligent Networks, 3*, 150–164.
[18] Kajol, R., & Akshay, K. K.. 2018. Automated agricultural field analysis and monitoring system using IOT. *International Journal of Information Engineering and Electronic Business, 11*, 17.
[19] Kaniki, A. M. 1988. Agricultural information services in less developed countries. *International Library Review, 20*, 321–336.
[20] Khan, H. H., Malik, M. N., Konečná, Z., Chofreh, A. G., Goni, F. A., &Klemeš, J. J. 2022. Blockchain technology for agricultural supply chains during the COVID-19 pandemic: Benefits and cleaner solutions. *Journal of Cleaner Production, 347*, 131268.
[21] Khattab, A., Abdelgawad, A., & Yelmarthi, K..2016. Design and implementation of a cloud-based IoT scheme for precision agriculture. In *2016 28th International Conference on Microelectronics (ICM)* (pp. 201–204). IEEE.
[22] Kizilaslan, N. 2006. Agricultural information systems: A national case study. *Library Review, 55*, 497–507.
[23] Klerkx, L., Jakku, E., & Labarthe, P. 2019. A review of social science on digital agriculture, smart farming and agriculture 4.0: New contributions and a future research agenda. *NJAS-Wageningen Journal of Life Sciences, 90*, 100315.
[24] Kobayashi, T., Tamaki, K., Tajima, K., Yoshimura, M., & Kato, M. 1990. A study for robot application in agriculture. *Journal of Agricultural Science, Tokyo Nogyo Daigaku, 35*, 80–87.
[25] Kontsevoy, G. R., Ermakov, D. N., Rylova, N. I., Leoshko, V. P., & Safonova, M. F. 2020. Management accounting of agricultural production: improving planning and standardization of costs in the management information system. *Amazonia Investiga, 9*, 284–293.
[26] Liu, S., Guo, L., Webb, H., Ya, X., & Chang, X. 2019. Internet of Things monitoring system of modern eco-agriculture based on cloud computing. *IEEE Access, 7*, 37050–37058.

[27] Liu, Y., Ma, X., Shu, L., Hancke, G. P., & Abu-Mahfouz, A. M. 2020. From Industry 4.0 to Agriculture 4.0: Current status, enabling technologies, and research challenges. *IEEE Transactions on Industrial Informatics, 17*, 4322–4334.

[28] Mahmud, M. S. A., Abidin, M. S. Z., Emmanuel, A. A., & Hasan, H. S. 2020. Robotics and automation in agriculture: Present and future applications. *Applications of Modelling and Simulation, 4*, 130–140.

[29] Mangla, S. K., Kazançoğlu, Y., Yıldızbaşı, A., Öztürk, C., & Çalık, A. 2022. A conceptual framework for blockchain-based sustainable supply chain and evaluating implementation barriers: A case of the tea supply chain. *Business Strategy and the Environment, 31*, 3693–3716.

[30] Mavilia, R., & Pisani, R. 2022. Blockchain for agricultural sector: The case of South Africa. *African Journal of Science, Technology, Innovation and Development, 14*, 845–851.

[31] Milovanović, S. 2014. The role and potential of information technology in agricultural improvement. *Economics of Agriculture, 61*, 471–485.

[32] Msoffe, G. E., & Ngulube, P. 2016. Agricultural information dissemination in rural areas of developing countries: A proposed model for Tanzania. *African Journal of Library, Archives & Information Science, 26*, 169.

[33] Mtega, W. P., & Msungu, A. C. 2013. Using information and communication technologies for enhancing the accessibility of agricultural information for improved agricultural production in Tanzania. *The Electronic Journal of Information Systems in Developing Countries, 56*, 1–14.

[34] Muangprathub, J., Boonnam, N., Kajornkasirat, S., Lekbangpong, N., Wanichsombat, A., & Nillaor, P. 2019. IoT and agriculture data analysis for smart farm. *Computers and Electronics in Agriculture, 156*, 467–474.

[35] Nayal, K., Raut, R. D., Narkhede, B. E., Priyadarshinee, P., Panchal, G. B., & Gedam, V. V. 2023. Antecedents for blockchain technology-enabled sustainable agriculture supply chain. *Annals of Operations Research, 327*, 293–337.

[36] Nithin, P. V., & Shivaprakash, S. 2016. Multi purpose agricultural robot. *International Journal of Engineering Research, 5*, 1129–1254.

[37] Ozdogan, B., Gacar, A., & Aktas, H. 2017. Digital agriculture practices in the context of agriculture 4.0. *Journal of Economics Finance and Accounting, 4*(2), 186–193.

[38] Paul, P. K., Ghosh, M., & Chaterjee, D. 2014. Information Systems & Networks (ISN): Emphasizing agricultural information networks with a case Study of AGRIS. *Scholars Journal of Agriculture and Veterinary Sciences. 1*, 38–41.

[39] Paul, P. K. 2015. Information and communication technology and information: their role in tea cultivation and marketing in the context of developing countries—A theoretical approach. *Current Trends in Biotechnology and Chemical Research, 5*, 155–161.

[40] Paul, P. K. 2016. Cloud computing and virtualization in agricultural space: A knowledge survey. *Palgo Journal of Agriculture, 4*, 202–206.

[41] Paul, P. K., Aithal, P., Sinha, R., Saavedra, R., & Aremu, B. 2019. Agro informatics with its various attributes and emergence: Emphasizing potentiality as a specialization in agricultural sciences—A policy framework. *IRA-International Journal of Applied Sciences, 14*, 34–44.

[42] P. K. Paul, Sinha, R. R., Baby, P., Shivraj, K. S., Aremu, B., & Mewada, S. 2020. Usability engineering, human computer interaction and allied sciences: With reference to its uses and potentialities in agricultural sectors: A scientific report. *Scientific Review, 6*, 71–78.

[43] Pedersen, S. M., Fountas, S., Have, H., & Blackmore, B. S. 2006. Agricultural robots—System analysis and economic feasibility. *Precision Agriculture, 7*, 295–308.

[44] Rao, N. H. 2007. A framework for implementing information and communication technologies in agricultural development in India. *Technological Forecasting and Social Change, 74*, 491–518.

[45] Reddy, P. K., & Ankaiah, R. 2005. A framework of information technology-based agriculture information dissemination system to improve crop productivity. *Current Science, 88*, 1905–1913.

[46] Rezník, T., Charvát, K., Lukas, V., Charvát Jr, K., Horáková, Š., & Kepka, M. 2015. Open data model for (precision) agriculture applications and agricultural pollution monitoring. In *EnviroInfo and ICT for Sustainability 2015*. Atlantis Press.

[47] Rose, D. C., & Chilvers, J. 2018. Agriculture 4.0: Broadening responsible innovation in an era of smart farming. *Frontiers in Sustainable Food Systems, 2*, 87.

[48] Tarannum, N., Rhaman, M. K., Khan, S. A., & Shakil, S. R. 2015. A brief overview and systematic approach for using agricultural robot in developing countries. *Journal of Modern Science and Technology, 3*, 88–101.

[49] Tsekouropoulos, G., Andreopoulou, Z., Koliouska, C., Koutroumanidis, T., & Batzios, C. 2013. Internet functions in marketing: Multicriteria ranking of agricultural SMEs websites in Greece. *Agrárinformatika/journal of agricultural informatics, 4*, 22–36.

[50] Vikram, P. R. K. R. 2020. Agricultural Robot–A pesticide spraying device. *International Journal of Future Generation Communication and Networking, 13*, 150–160.

[51] Yan-e, D. 2011. Design of intelligent agriculture management information system based on IoT. In *2011 Fourth International Conference on Intelligent Computation Technology and Automation* (Vol. 1, pp. 1045–1049). IEEE.

[52] Zhai, Z., Martínez, J. F., Beltran, V., & Martínez, N. L. 2020. Decision support systems for agriculture 4.0: Survey and challenges. *Computers and Electronics in Agriculture, 170*, 105256.

[53] Zhang, Y., Wang, L., & Duan, Y. 2016. Agricultural information dissemination using ICTs: A review and analysis of information dissemination models in China. *Information Processing in Agriculture, 3*, 17–29.

5 Smart Agriculture Based on Artificial Intelligence in the African Region
Open Challenges, Solutions

Ametovi Koffi Jacques Olivier, Taushif Anwar, Ghufran Ahmad Khan, and Zubair Ashraf

5.1 INTRODUCTION

In Africa, agriculture constitutes the predominant economic activity and occupies a place of the highest importance. It forms the core of economic operations in many developing countries, propelling industrial progress and facilitating the transformation of these economies. Furthermore, agriculture plays a multifaceted role in development, encompassing various aspects of the process, including economic revitalization, job creation, and contribution to the value chain, poverty reduction, and reduction of inequality of income, ensuring food security, providing environmental benefits and generating foreign capital.

The data presented in Figure 5.1 illustrates the distribution of economic sectors within the gross domestic product (GDP) of specific regions in 2021. Unsurprisingly, agriculture emerges as the leading sector in Africa, accounting for 17.27% of raw domestic product. This sector also employs around two-thirds of the continent's workforce and contributes between 30 and 60 percent to the GDP of African countries. It is important to note that the majority of African farmers are smallholders; however, their farms have lower productivity than other developing regions, perpetuating rural poverty across the continent. The challenges facing African agriculture include rapid population growth, land degradation, climate change and fluctuating food prices, all of which are putting increasing pressure on agricultural production and food security in Africa.

Africa has a higher agricultural ratio than some regions of the world; its agricultural productivity rate was 17.24% according to the 2021 statistics in Figure 5.1, compared with 22.54% in the rest of Asia (South Asia, East Asia and the Pacific). This shows the serious situation of the African agricultural sector. In addition to this, agriculture has been neglected in recent years due to various factors such as the privatization of the agricultural sector or the limited resources of the sector, because

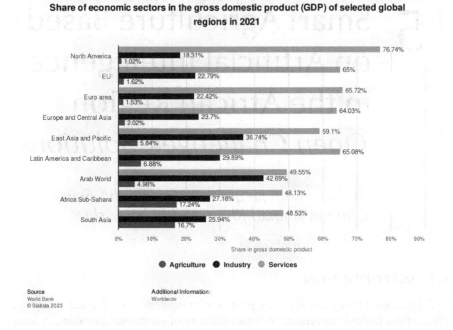

FIGURE 5.1 Share of economic sectors in the gross domestic product by global regions 2021. (World Bank.)

while the expansion of cultivated land has allowed in the past a significant increase in agricultural production, this is not the case today. It is necessary to move from so-called traditional agriculture to modern agriculture, using satellites to provide precise climate data, to cutting-edge innovations such as the Block Chain, including smartphones, artificial intelligence and connected objects [5].

These technologies can be a game-changer, making agriculture more productive and resilient in a sustainable way to feed a world of 9 billion people by 2050 [3], [6]. In this study, we review the challenges facing agriculture in Africa. We will briefly describe the issues currently facing sub-Saharan Africa, starting with Ivory Coast's various export commodities, including oil palm, cocoa and rubber. We will then present recommendations to resolve these issues in agriculture so that Africa can become a true economic powerhouse through profitable agriculture.

5.2 LITERATURE REVIEW ON MODERN AGRICULTURE

Artificial intelligence (AI) solutions are not just theoretical concepts; they are practical tools that boost efficiency in various disciplines and address difficulties encountered by numerous industries, including agriculture. They are tackling issues such as yield of crops, irrigation, soil composition evaluation, crop monitoring, weed management, and planting of crops (Kim et al. [7]). Agricultural Robots, a tangible application of AI, are being used to enable high-quality uses in this industry. As the world's population grows, the agriculture industry is facing significant challenges, but AI has the potential to provide much-needed solutions. AI-based technical solutions are not just promises; they already allow farmers to produce more with less, improve output quality, and ensure that harvested products reach the market sooner. By 2020, farmers will be using 75 million linked gadgets. By 2050, the ordinary farm will no longer exist, with an average of 4.1 million data points created daily. Here are some of the practical ways that artificial intelligence is benefiting the agriculture industry:

5.2.1 Digital Agriculture Transformation

According to research by Ranveer Chandra et al. [6], small producers' productivity, transparency, and financial success may all be considerably increased by using digital technologies like AI, CV (Computer Vision), and the IoT (Internet of Things). Existing artificial intelligence and computer vision systems, such as Farmers Edge, Climate Field View, and Land O'Lakes R7, may predict weather, yields, outcomes, crop stress, and more. Provide farmers with speech and natural language interfaces.

5.2.2 The Impact of AI in Agriculture

Matthew J. Smith [8] discusses many areas where AI will add value, one of which remains agriculture. In his view, artificial intelligence will provide more and better information on farm conditions, allowing farmers to learn, think, and decide on management measures, and learn, think, and decide on management measures. For example, artificial intelligence is to help farmers get the information they need in the right way, and let farm workers translate the information they want into the most suitable language for them, and provide it to farmers working in different language areas to help agriculture suggestion. These skills can significantly improve the dissemination of good agricultural advice, allowing better agricultural practices to be more widely promoted

5.2.3 Recognition of Images

According to Lee et al. [9], interest has risen over the past several years in Recognition of images and awareness about them. It has been noticed in autonomous drones and their applications, including detection and surveillance, person identification and localization, search and rescue, and wildfire detection (Bhaskaranand & Gibson [10] and Doherty & Rudol [11]). In addition to being able to travel enormous distances and heights, drones, also known as unmanned aerial vehicles (UAVs), are becoming increasingly popular and are of tremendous significance (Tomic et al. [12] and Merino et al. [13]).

5.2.4 Skills and Manpower

Panpatte [14] says AI makes this possible Farmers need to collect a lot of data from both government and government public website that analyses everything and offers solutions to farmers. It solves a lot of unclear problems and gives us a smarter way of irrigation, resulting in higher yields for farmers. Due to artificial intelligence, agriculture will be a combination of science and technology In the near future, biological capabilities will not only improve quality results for all farmers, but also minimize them their loss and workload. Two-thirds of the world's population will disappear by 2050; UN says the population will live in urban areas, which means a reduction in the need for burden on farmers. Applications of AI in Agriculture Will automate multiple processes, reducing risk and benefiting farmer's relatively simple and efficient farming.

5.2.5 Maximize Productivity

Ferguson et al. [15] identify the maximum performance level of any system in their Pot as Variety Selection and Seed Quality. The advancement of technology has aided in crop selection and the identification of the best hybrid seeds for landowners' requirements. This is accomplished by knowing how the seed is put into action. Address various environmental variables and soil kinds. Acquiring this data can help to lower the risk of plant infections. We can now match trends in the market, annual performance, and customer requirements, allowing farmers to optimize agricultural yields with greater efficiency.

5.2.6 Farmer ChatBot

Chatbots are just artificial conversation robots that automate conversations with consumers. Conversational artificial intelligence and automated learning technologies enable us to interpret spoken language and connect with consumers in a more personalized manner. They are primarily employed in retail, tourism, and media; agriculture has taken advantage of this chance to assist farmers in obtaining answers to their lingering problems, as well as to give advice and ideas.

5.2.7 Total Non-involvement of African States

Africa has not yet understood the immense power that its agriculture can wield if properly developed. The major cause is the total non-involvement of all African states in the immense assets of the sector. African heads of state do not understand that modern, technological agriculture like that practiced in Asian countries could exponentially improve the financial health of African nations. The real key to the progress of African nations is the industrialization of their traditional agriculture and this is only possible because of the urgency of product processing.

In a report published by the Food and Agriculture Organization of the United Nations, the yield of agricultural plantations must increase by around 70% over the next few years to be able to meet the needs of the populations of Africa. To be able to meet the needs of African populations through agricultural yields, we must address challenges that nevertheless remain ignored by political decision-makers [4].

African agriculture is still practiced in a traditional way, despite the enormous power it can unleash if properly developed. The main reason is the total non-participation of

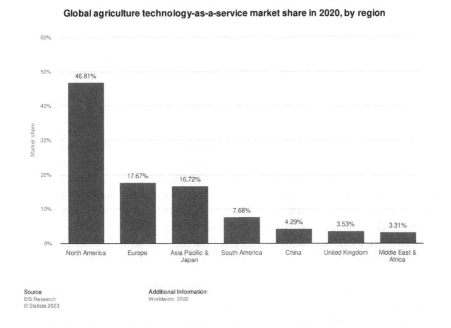

FIGURE 5.2 Global agriculture technology-as-a-service market share in 2020, by region. (BIS Research.)

all African countries in this industry which constitutes a considerable asset because even if African heads of state know that agriculture is the engine of their economy, the fact of not opting for modern technological agriculture like that practiced in Asian countries this is delaying the sector as shown in (Figure 5.2) which shows the Global Agricultural Technologies as a Service Market Share in 2020, by region

In 2020, North America dominated the agricultural technology services market, accounting for 46.81% of the total market, followed by Europe (17.67%), Asia Pacific and Japan (16.72%). As for Africa, it comes at the bottom of the pack with 3.31%. Faced with these results, a report published by the Food and Agriculture Organization of the United Nations indicates that the production of agricultural plantations must increase by around 70% over the coming years to meet the needs of African populations. To be able to meet the needs of the African population through agricultural production, we must address the challenges that remain ignored by political decision-makers [4].

5.2.8 Farmers without Training on Good Practice Methods

The lack of training and practical advice is also a problem facing African farmers. Many have received no training or practical advice on the techniques and methods associated with modern agriculture. Unfortunately, a large number of farmers in Africa are uneducated and live in villages and camps with limited access to telecommunications services such as television, radio and even the Internet. There are also people who do not have access to good-quality seeds. Research on this issue shows that some seeds are not very resistant, depending on climatic conditions and the use of agricultural land. Good seed is very expensive and in most cases farmers cannot differentiate between them, which could result in poor yields [4].

5.2.9 THE HIGH PRICE OF FERTILIZERS AND PESTICIDES

African land is extremely coveted by agro-industries, driving up fertilizer prices across the country, thus preventing small farmers from having access to fertilizer based on their financial situation, hence the use of the same land endlessly. Mishandled, land degrades and loses its components necessary for good agricultural production. To deal with this situation, farmers are using artificial fertilizers as their main method in order to work the same land as often as possible. As these (chemical) fertilizers are quite expensive there, they are out of reach of farmers or unavailable in rural markets. In countries where fertilizer subsidies exist, part of the rural population is excluded from distribution [4].

As proof, Figure 5.3 shows that in 2022, the world's largest fertilizer consumer will be in East Asia, accounting for around a third of global consumption. Overall, Asia accounts for about half of global fertilizer consumption. Meanwhile, the Americas (which include North America, Latin America and the Caribbean) follow with a good percentage and Africa remains at the back of the pack with 3%, proving that the problem is indeed worrying for farmers.

Much like the distribution of pesticide use worldwide, according to FAO data (Figure 5.4), the Americas will account for more than half of global agricultural

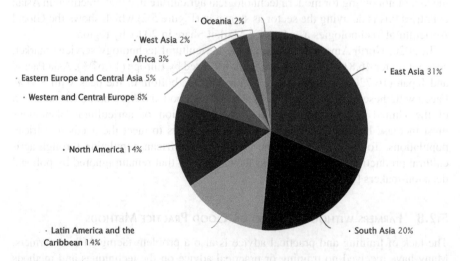

FIGURE 5.3 Distribution of Agriculture Fertilizer Consumption Worldwide in 2022, courtesy: International Fertilizer Industry Association, Nutrien in 2022.

AI in African Agriculture: Challenges and Solutions

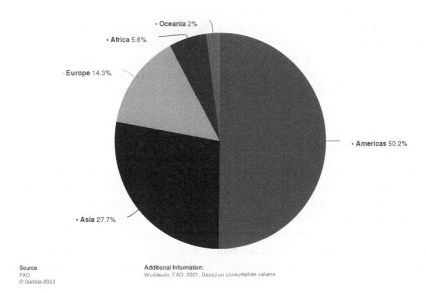

FIGURE 5.4 Global pesticide use distribution 2021, by region. (FAO.)

pesticide consumption in 2021. Asia and Europe follow with shares of 27.7% and 14.3% respectively. Africa and Oceania represent less than 8% of total pesticide consumption in the world, which could explain low yields due to the difficulty of easily obtaining the products necessary to increase their production.

5.2.10 Limited Access to Processing Facilities

The lack of industrialization and limited availability of storage facilities in Africa pose a significant challenge for farmers in rural areas. This leads to production losses as crops cannot be efficiently transported to processing plants, mainly due to poor or even lack of road networks. In order to transport agricultural products to markets, well-maintained roads are essential. However, either these roads do not exist, or those which connect the villages to the main markets are in poor condition. As a result, small producers experience reduced yields, preventing them from achieving financial stability. The agricultural sector in Africa faces a major obstacle due to these circumstances [1], [4].

5.2.11 Difficulties in Accessing Financing and Credits

African governments do not provide financial support institutions to small plantation owners to improve their agricultural profitability. Despite the existence of some microfinance and banks, access to financing for small planters remains a real obstacle, because planters in rural areas are sometimes unable to meet the eligibility conditions due to their precarious financial situation or their inability to deliver their production. Due to the long waiting time before payment once delivered to their planter number at the factory. It should be noted that access to credit in Africa differs from that in Asian countries, where financial institutions and the state are actively involved in regulating the market to promote a better life for small farmers by providing them with credit in according to their respective needs [2], [4].

5.3 INDUSTRIAL AGRICULTURE AND ITS CHALLENGES

Industrial or export agriculture involves the cultivation of plants of high economic value for commercialization. It was introduced to the African continent (North, South, East, Central and West Africa) by settlers and practiced in agro-industrial, medium and small areas that produce agricultural products such as oil palm, cocoa, rubber and other agricultural products. This section will provide an overview of some African cash crops, such as oil palm, rubber and cocoa, the highest production of which is in West Africa: Ivory Coast, one of the African leaders in the export of staple crops (Figure 5.5).

5.3.1 Challenges of oil palm industry

Africa, the birthplace of oil palm, is at the center of all the aspirations of agri-food multinationals, because palm oil is at the heart of the traditions and life of many communities around the world, since it represents more than 50% of global consumption. Vegetable oils are not only used in cosmetics and agricultural products, but are also an important part of the continent's economy and food security, but the production of this product faces many challenges. Currently, it is not one of the main world

FIGURE 5.5 PalmElit: CIRAD® oil palm seeds.

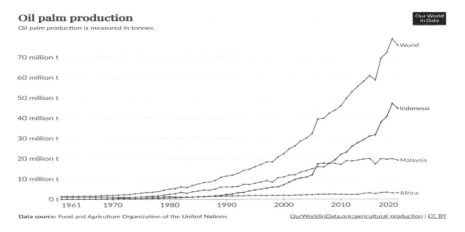

FIGURE 5.6 Oil palm production. (Food and Agriculture Organization of the United Nations.)

producers, which are Malaysia and Indonesia (Figure 5.6), which hold the majority of production even though their seeds come from Africa.

How can we explain this turn of events? This is likely a result of major controversies in the industry, including:

- **Deforestation:**

Although deforestation is often necessary to establish oil palm cultivation, it also endangers the biodiversity of producing countries such as the Ivory Coast, where the country's elephant is disappearing.

- **Modernization and industrialization:**

Poor infrastructure and market access difficulties due to poor road network and remoteness of palm oil processing plants. Most farmers also use fire for land preparation because they do not have the financial resources to purchase labor or necessary mechanical tools, showing that African agriculture is still practiced traditionally compared to that of Asian countries, which is practiced with precision technologies and powered by tractors, etc. as demonstrated in Figure 5.7. When comparing with other agricultural items that were imported, the percentage of technology for agriculture that was brought into West Africa in the year 2020 that comprised of bulldozers and excavators was the most significant percentage. In addition, 13 percent of the overall farm technology imports in the region were comprised of agricultural chemicals, while 12 percent consisted of insecticides, herbicides, and other items of the same sort.

- **Monitoring and authentication of production:**

The capacity of manufacturers to guarantee the origin of palm bunches, by plot and by producer, is not yet sufficiently effective due to the variety and quality

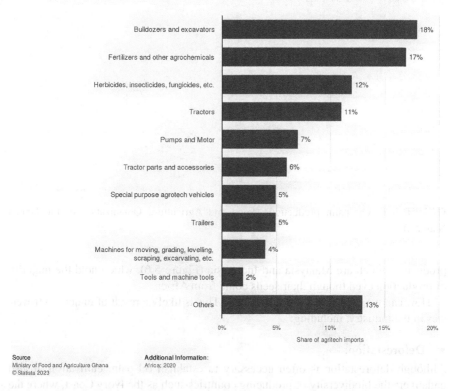

FIGURE 5.7 Share of agricultural technology imports into West Africa as of 2020, by product. (Ministry of Food and Agriculture Ghana.)

of the fruits they grow, not favoring closer coordination between the different players in the sector. It is also essential that the implementation of all stages of the process, such as processing, transportation, manufacturing and marketing, can be monitored and verified by technological solutions such as an application [6], [22].

- **Poor nutritional status of palm groves:**

The nutritional need of oil palm plantations corresponds to the quantity of the nutrient necessary and sufficient to achieve a high production objective. The absence or low level of these nutrients such as potassium, nitrogen, calcium, etc. limits the yields of small producers who do not apply fertilizers due to high costs, lack of knowledge about the signs of deficiencies and lack of financial support from institutions. Some plantations have never received any nutrition during the last 10 years or since the start of planting, which is insufficient, especially if the contributions are not made every year or on all surfaces [4], [6] (as shown in Figure 5.8).

AI in African Agriculture: Challenges and Solutions

FIGURE 5.8 Rubber – Tapping, Coagulation, Processing | Britannica. (https://images.app.goo.gl/udwc1Jug1Xga7xkT7)

5.3.2 CHALLENGES OF RUBBER INDUSTRY

Besides the oil palm, Africa has another wealth: the rubber tree. Better known as the rubber tree because of its origin, this fast-growing tropical tree is cultivated and planted in the Amazon, West Africa, and especially in Southeast Asia. The first African place is occupied by Côte d'Ivoire, which aims to rise to first place in the world in the coming years. According to information from the CIA (World Factbook) and WTEx (Figure 5.9), natural rubber is Africa's second largest agricultural export product, of which Ivory Coast ranks third in global natural rubber exports 2022 with 11.6%, just behind Indonesia with 22% and Thailand around 31.6%, making it the world's largest producer of natural rubber. Africa despite its current ranking could achieve better results, but the rubber industry faces enormous challenges, starting with:

- **Workforce:** Labor is difficult to find everywhere in many countries, This is a very difficult problem for rubber growing, especially for the plantation operator, because he must find employees responsible for tapping, and these employees must be familiar with the delicate operations of harvesting latex rubber tree. But unfortunately, due to various factors such as the instability of global rubber prices, labor has become scarce, which has made labor resources difficult to access as this labor force work turned to the exploitation of gold, which they considered more profitable and more stable [27].

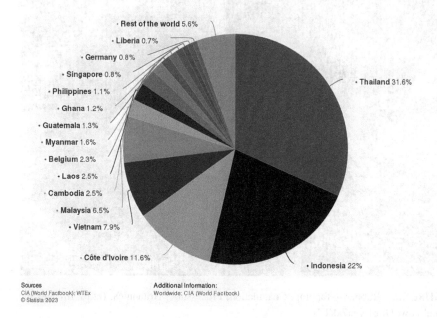

FIGURE 5.9 Distribution of natural rubber exports based on value in 2022, by country. (CIA (World Factbook); WTEx)

- **Rubber growing techniques:** Rubber growers derive their income from dry rubber content (DRC), rather than from the amount of raw rubber they produce. Not surprisingly, rubber growers are successful in adopting techniques that maximize long-term yields of dry rubber, while rubber growers in rural areas with little knowledge of these techniques often cannot afford to adopt them, reducing their productivity [28].
- **Rubber tree diseases:** The main problem in rubber cultivation in Africa is parasitic diseases, which can cause secondary rubber leaf fall. It is caused by an Ascomycota Melanconiale fungus called Colletotrichum gloeosporioides. Secondary rubber tree leaf drop occurs as the tree replenishes its leaves. This sensitive phase of its phenology occurs between March and April, during the long rainy season which favors fungal growth. This causes severe attacks on susceptible clonal plants, resulting in necrosis and deformation of the leaves, followed by the loss of young leaves affected in the early stages of development which are systematically destroyed by the attacks these fungi and the leaves are less and less vigorous as they follow one another, which prevents the trees from reconstituting their initial foliage. The most vigorous plants can only return to their normal leaf density between June and July. It is therefore necessary to find aerial treatments based on products that cause defoliation and re-foliation before conditions favorable to the development of the fungus are established in order to eradicate it completely [30].

FIGURE 5.10 Varieties of Cocoa Beans | Learn Chocolate – Bar & Cocoa.

- **The genomic choice of the rubber tree:** Rubber tree varieties grown by African producers are generally grafted clones, planted at densities of 400 to 500 trees per hectare and harvested at around 5 to 7 years old. As these clones are bled, they break from wind, succumb to disease, or stop producing due to changes in their milk cells, a phenomenon known as "stem notching." It is therefore necessary to provide genetically modified plants to farmers in village areas to support the development of cultivation, taking into account increasingly diversified objectives (increasing the profitability of farms, perpetuating the production of natural rubber) and by integrating an innovative range of rubber genomes. Some improved rubber plants exist, but the high costs prevent small producers from using them to have better production and fight against diseases and insects that tire rubber crops [23-26] (as shown in Figure 5.10).

5.3.3 Challenges of Cocoa Industry

Tropical equatorial regions are suitable for cultivating cocoa plants and beans. Even though cocoa beans are from South America, Ivory Coast is the world's most significant cocoa producer and exporter. According to the Coffee Cocoa Council (CCC), the production of cocoa beans in the region in 2022 was 2,358,840 tonnes, ranking it 1st in the world. However, in this African country, cocoa mobilizes 1 million producers and provides income for 20% of the population. Cocoa is the country's primary foreign exchange earner, accounting for around 14% of GDP and around 40% of exports, and is, therefore, an essential source of income. Still, it is fragile due to many challenges, such as:

- **Child labor:** Child labor is defined by the International Labor Organization as "work that deprives children of their childhood, their potential and their dignity, and that harms their physical and mental development." Forced child labor is a risk present in the cocoa supply chain, although it is generally less common. When children are taken from their families, with or without their consent, to be exploited in cocoa production, this is a case of human trafficking and represents a complex problem for a more dynamic sector to resolve [16].

- **Problem of diseases such as swollen shoots:** Swollen shoots disease is caused by at least ten species of viruses of the genus Badnavirus belonging to the Caulimoviridae family. These viruses are only observed on the cocoa tree in West Africa, not in Latin America, the region of origin of the cocoa tree, and it is likely that they were transmitted to the cocoa tree from wild host plants of the family Malvaceae. In infected plant organs, the virus is mainly present in the phloem companion cells and causes tissue growth disorders and impedes sap circulation, leading to the death of cocoa trees within a few years [18].
- **Competition from Asian countries that use modern technological agriculture:** Asia is certainly the youngest cocoa-producing region, but has great economic potential. Recent innovations in planting, production and processing techniques have elevated it to the ranks of potential competitors in Africa's cocoa sector. These Asian countries are arming themselves with new techniques, new programs and have the financial support of the American administration, as well as the technical support of agri-food multinationals like Cargill in order to constantly improve the quality of production, which unfortunately is not what it is today. This is not the case for African cocoa-producing countries [19].
- **Cocoa quality and packaging:** Large-scale deforestation, the increased impact of long dry seasons, the impact of harmattan on cocoa trees, the aging of plantations and the reduction in tree resistance, as well as the decline in prices discouraged producers (and their workers), leading to a drop in pesticide treatments, which also had an impact on the cocoa tree. The health of the pods and the health of the cocoa beans. Increased competition between trackers and buyers who do not care about quality, but only quantity has led to an increasing number of producers neglecting the sorting of cocoa pods and beans, leading to a decline in cocoa quality . Furthermore, Decree No. 2012.1011 defines the rules related to the packaging of cocoa intended for export such as correct fermentation; maximum humidity of 8%; cleanliness, the absence of free or adhering foreign bodies (fragments, mineral components) and the absence of any odor, in particular musty, smoke or pesticides. Due to poor technical and material conditions in rural plantations, the rules cannot be respected, thus creating obstacles to the sector [17].

The oil palm, rubber and cocoa industries, as well as other export crops, face common problems such as lack of access to fertilizers, lack of modernization and lack of real structured financial support for producers. But the most worrying issue remains: the absence of fair trade tariffs by multinational companies, the narrowness of regional markets and the overcapacity of industrial units, the urgency of increasing the possibilities for local processing of raw material derivatives and the need for greater regulation of the agricultural sector in each country. Faced with these numerous problems in these various industries, our next section will be devoted to practical recommendations that could be used to resolve them.

5.4 PRACTICAL RECOMMENDATIONS

Regarding the practical recommendations section, it is important to emphasize that despite the challenges facing the African continent, it occupies a significant global position that could be improved and change the living conditions of producers, thus contributing to the wealth of its growing population, if they adopt strategies such as:

- Modernize and industrialize the different sectors, which would involve increasing the capacity for transformation and diversification of raw materials into semi-finished and finished products, requiring investments;
- Optimize value chains by seeking to conquer new markets and potential partnerships with the largest consumers of African products such as India, China, etc.;
- Improve the structure of the sectors to increase the income and conditions of farmers;
- Regulate activities by promoting and certifying the quality of production.

In addition to these initial recommendations, it is essential to overcome the difficulties of integrating modern technologies into African agriculture. Emerging technologies such as Big Data, Internet of Things, and Artificial Intelligence can revolutionize African agriculture by improving precision solutions. For example, the use of drones can help detect and monitor crop stress. Additionally, an artificial neural network can analyze agronomic data collected by applications. Hyper spectral remote sensing can also be used to study agricultural pests and diseases. It is crucial to recognize that African agriculture has the potential to become a global force and thrive through the use of technologies similar to those used in Asian countries such as Malaysia. For example, the Sustainable Palm Oil Cluster (SPOC) is used to map and develop applications to promote Good Agricultural Practices (GAP) among smallholders in Malaysia (as shown in Figure 5.11).

This is why we propose a new framework that focuses on the main aspects of the economic life of African farms, in particular the factors of production. By investing in these emerging technologies and adopting sustainable agricultural practices, African agriculture can not only increase its productivity but also improve the quality of its products. This will enable African agriculture to position itself on the global market and contribute to global food security. It is time to seize this opportunity and implement policies and programs that promote the adoption of these technologies in African agriculture. Together, we can create a prosperous future for African agriculture and its farmers.

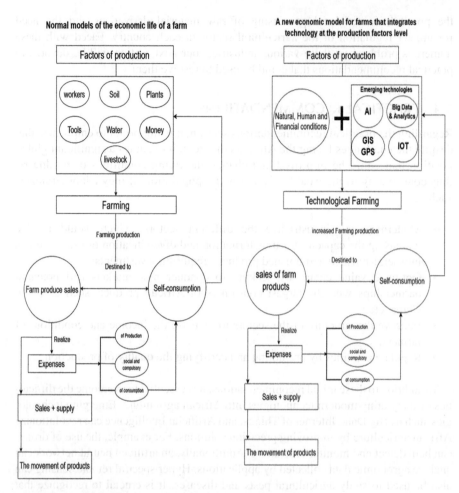

FIGURE 5.11 Normal model and New economic model for form.

5.5 CONCLUSION

In this article, we discuss the difficulties faces in various agricultural sectors in the African region. Every day, new technological and digital solutions are developed to increase the yield of agricultural production. Therefore, we provided a possible solution to modernize its agriculture by using digital technologies such as Artificial Intelligence (AI), the Internet of Things (IoT), and block chain. As a result, African scientists will need to implement technological models and tools to help the transition from traditional to digital agriculture with the aim of improving Africa's situation. Our future studies will focus on the implementation of agricultural solutions to help Africa progress towards digital agriculture.

REFERENCES

[1] NEPAD. (2013). Les agricultures africaines, transformations et perspectives, NEPAD, Novembre 2013, 72 p.

[2] Exploiter le Potentiel Agricole pour la Croissance et le Développement en Afrique de l'Ouest. Commission Economique pour l'AfSrique CEA, 2012.

[3] Adesina, A., President, African Development Bank, October 2015. Unlocking Africa's potential to create wealth from agriculture. www.afdb.org/en/news-and-events/unlocking-africas-potential-to-create-wealth-from-agriculture-14841

[4] Isidore Kpotufe of IMANI, the Africa and the World team. Le secteur agricole africain: Les défis à relever. www.hubrural.org/Le-secteur-agricole-africain-Les.html?lang=en

[5] African Union. Agricultural Development. https://au.int/en/agricultural-development.

[6] Chandra, R., Collis, S., 2021. Digital agriculture for small-scale producers: challenges and opportunities. *Commun. ACM*, 64(12), 75–84. https://doi.org/10.1145/3454008

[7] Kim, Y.J., Evans, R.G., Iversen, W.M., 2008. Remote sensing and control of an irrigation system using a distributed wireless sensor network. *IEEE Trans. Instrum. Meas.* 57(7), 1379–1387. https://doi.org/10.1109/TIM.2008.917198

[8] Smith, Matthew J., 2018. Getting value from artificial intelligence in agriculture. *Animal Prod. Sci.* 60 (1), 46–54.https://doi.org/10.1071/AN18522

[9] Lee, J., Wang, J., Crandall, D., Sabanovic, S., Fox, G., 2017. Real-time, cloud-based object detection for unmanned aerial vehicles. *2017 First IEEE International Conference on Robotic Computing (IRC)*. https://doi.org/10.1109/irc.2017.7

[10] Bhaskaranand, M., Gibson, J.D., 2011. Low-complexity video encoding for UAV reconnaissance and surveillance. *Proceeding of the IEEE Military Communications Conference (MILCOM)*, pp. 1633–1638. https://doi.org/10.1109/MILCOM.2011.6127543

[11] Doherty, P., Rudol, P., 2007. A UAV search and rescue scenario with human body detection and geolocalization. In: Orgun, M.A., Thornton, J. (eds.), *AI 2007: Advances in Artificial Intelligence. AI 2007. Lecture Notes in Computer Science*. Vol 4830. Springer, Berlin, Heidelberg, pp. 1–13. https://doi.org/10.1007/978-3-540-76928-6_1.

[12] Tomic, T., Schmid, K., Lutz, P., Domel, A., Kassecker, M., Mair, E., Grixa, I.L., Ruess, F., Suppa, M., Burschka, D., 2012. Toward a fully autonomous UAV: Research platform for indoor and outdoor urban search and rescue. *IEEE Robot. Automation Magazine* 19 (3), 46–56. https://doi.org/10.1109/mra.2012.2206473

[13] Merino, L., Caballero, F., Martínez-de Dios, J.R., Ferruz, J., Ollero, A., 2006. A cooperative perception system for multiple UAVs: Application to automatic detection of forest fires. *J. Field Robot.* 23 (3–4), 165–184. https://doi.org/10.1002/rob.20108

[14] Panpatte, D.G., (2018). *Artificial intelligence in agriculture: An emerging era of research*. Anand Agricultural University: 1–8.

[15] Ferguson, R.B., Shapiro, C.A., Hergert, G.W., Kranz, W.L., Klocke, N.L., Krull, D.H., 1991. Nitrogen and irrigation management practices to minimize nitrate leaching from irrigated corn. *Jpa* 4 (2), 186. https://doi.org/10.2134/jpa1991.0186

[16] Child labor in cocoa. www.cocoainitiative.org/issues/child-labour-cocoa

[17] Ivorian cocoa: Standards to be respected for export. https://castor-ci.com/en/le-cacao-ivoirien-les-normes-a-respecter-a-lexportation/

[18] Swollen shoot. https://barco.cirad.fr/projet/swollen-shoot.

[19] The provenance of cocoa beans from Asia. https://chococlic.com/La-provenance-des-feves-de-cacao-d-Asie_a1516.html.
[20] Côte d'Ivoire: National cocoa exporters denounce the monopoly of multinationals, 2021. www.koaci.com/article/2021/02/02/cote-divoire/economie/cote-divoire-les-exportateurs-nationaux-de-cacao-denoncent-le-monopole-des-multinationales_148 522.html.
[21] The consequences of the misuse of chemical fertilizers and pesticides. www.slateafrique.com/213133/les-consequences-de-la-mauvaise-utilisation-des-engrais-chimiques-et-pesticides.
[22] Bessou, C., Dubos B., 2020. Filière Palmier à Huile en Côte d'Ivoire: Analyse fonctionnelle et diagnostic agronomique, Rapport d'étude Cirad n°2912, réalisé pour le FIRCA et l'AIPH, 42p. Août 2020, Montpellier, France. https://agritrop.cirad.fr/597525/.
[23] Benoist, A., Leconte, A., 2020. Filière Hévéa en Côte d'Ivoire: Analyse fonctionnelle et diagnostic agronomique, Rapport d'étude Cirad, réalisé pour le FIRCA et l'APROMAC, 48p. Août 2020, Montpellier, France. https://agritrop.cirad.fr/597523/
[24] Bockel, L., Ouedraogo, S.A., Auguste, K.A., Gopal, P. 2021. *Analyse prospective de la filière cacao en Côte d'Ivoire 2020-2030 – Vers une politique commune de marché de cacao en Afrique de l'Ouest.* Accra, FAO. https://doi.org/10.4060/cb6508fr
[25] Maxime Cumunel. *Note 13 – February 2020.The palm oil sector in Côte d'Ivoire: a summary of the challenges of sustainable development.* The Foundation for World Agriculture and Rurality (FARM). https://fondation-farm.org/la-filiere-palmier-a-huile-en-cote-divoire-un-condense-des-enjeux-du-developpement-durable/.
[26] Oumar NDIAYE, Saïba FAINKE. 2021. Société Africaine de Plantations d'Hévéa (SAPH): ANALYSE Mai 2021.WARA.
[27] Bineta Diagne. 2022. Ivory Coast: the rubber tree and its paradoxes. Chronicle of raw materials. https://rfi.my/8rEF
[28] François RUF. « The Adoption of Rubber in Côte d'Ivoire. Prices, Copying effect, Ecological and social Change », Économie rurale, 330-331 | juillet-septembre 2012. Open Edition Journals. https://doi.org/10.4000/economierurale.3527
[29] Baudelaire Mieu. Côte d'Ivoire: la filière hévéa sous pression. Octobre 2013. www.jeuneafrique.com/15887/economie/c-te-d-ivoire-la-fili-re-h-v-a-sous-pression/
[30] Guide du conseiller agricole hévéa- tome 4: maladies et ravageurs de l'hévéa. Edition 2013. FIRCA. https://firca.ci/wp-content/uploads/2022/07/guide-conseiller-agricole-hevea-tome-4.pdf

6 Synergizing Artificial Intelligence and Optimization Techniques for Enhanced Public Services

Kapil Saini, Ravi Saini, and Parul Sehrawat

6.1 INTRODUCTION

Artificial Intelligence (AI) is revolutionizing public services by bringing about unparalleled creativity and efficiency. This introduction examines the diverse impact of AI on public services, including its development, effects, and the complex dynamics it brings to government operations. Recently, the convergence of AI and public services has become a central topic of discussion, propelled by a combination of technological progress and the need to update bureaucratic structures. As governments throughout the world struggle to provide services in a more efficient and responsive manner, Artificial Intelligence (AI) has emerged as a promising solution. The integration of artificial intelligence (AI) in public services has resulted in substantial enhancements in both operational efficiency and overall effectiveness. Artificial intelligence (AI) technologies, such as machine learning, natural language processing, and robotic process automation, are being used to make administrative procedures more efficient, improve decision-making, and increase service delivery. AI-driven chatbots and virtual assistants are transforming customer service by delivering immediate and precise answers to citizen queries, resulting in shorter wait times and higher levels of satisfaction. Furthermore, artificial intelligence's ability to make predictions is being utilized to anticipate and proactively answer the demands of the public. Predictive analytics may be utilized in various domains, including healthcare, where artificial intelligence algorithms analyze data to anticipate the occurrence of disease outbreaks. This enables prompt interventions and efficient allocation of resources. AI is used in law enforcement to predict and prevent crimes, improving public safety by using tactics based on data. The use of AI also brings about a significant change in the composition of the workforce in public sectors. AI streamlines mundane and repetitive processes, enabling human workers to dedicate their attention to intricate and strategic activities that need human judgement and creativity. In order to adapt to a workplace where artificial intelligence is incorporated, it is necessary to implement efforts that focus on acquiring new skills and enhancing existing ones.

Nevertheless, the incorporation of artificial intelligence in public services is not devoid of obstacles.

Data privacy, algorithmic prejudice, and the digital divide present substantial challenges. To guarantee the ethical deployment of AI, it is necessary to establish strong frameworks and rules that protect the rights of citizens and encourage transparency and responsibility. Moreover, the long-term viability of AI capabilities in public organizations generally relies on consistent investment and support. As previously mentioned, the absence of stable and consistent resources could make AI capabilities vulnerable and short-lived, underscoring the importance of strategic planning and enduring commitment from government entities.

6.1.1 The Morphing Landscape of Public Services

The public sector, known for its bureaucratic procedures and hierarchical organization, is currently experiencing a significant change due to the integration of artificial intelligence (AI) technologies. This revolutionizes governance from a conceptual to a technological level, placing greater emphasis on agility, responsiveness, and citizen-centricity. The transformative capabilities of AI transcend various domains within the public service sector, including policy formulation, decision support, and optimization.

6.1.2 A Symphony of Technological Advancements

The incorporation of artificial intelligence into public services is not a one-time occurrence but rather an ongoing symphony of technological advancements. As governments make use of machine learning algorithms, natural language processing, and predictive analytics, the very fabric of public administration is being rewoven. The orchestration of these technologies brings about a crescendo of efficiency, transparency, and innovation.

6.1.3 Open Innovation in Public Services

An examination of open innovation in the public sector through a systematic literature review unveils an intriguing fabric of collaborative endeavours, in which artificial intelligence (AI) serves as a catalyst for innovative governance approaches [1]. This dynamic interaction between AI and open innovation represents a deviation from conventional models, promoting an ecosystem that exchanges insights and addresses challenges collectively.

6.1.4 The Digital Era and Policy-making

An analysis of open innovation in the public sector conducted via a systematic literature review reveals a captivating network of collaborative efforts, with artificial intelligence (AI) acting as a catalyst. The impact of AI on policy-making is significant, surpassing conventional limitations and introducing a data-centric aspect to the decision-making process. Equipped with AI-generated insights, governments can adeptly navigate the intricacies of policy-making, guided by informed judgement [2].

6.1.5 LEADERSHIP IN THE AGE OF AI

Without strong digital and imaginative leadership, it will be impossible to fully exploit the revolutionary potential of artificial intelligence in public services [3]. In order to ensure that AI is a good force for change, governments need to find a balance between permitting innovation and implementing regulations.

6.2 PERSONALIZATION OF SOCIAL SERVICES

Artificial Intelligence (AI) has become a transformative force in education, offering unprecedented opportunities to identify and cater to individual differences among learners. In the pursuit of educational excellence, understanding and addressing these differences are paramount. This section explores how AI, with its advanced capabilities, facilitates the identification of individual differences, paving the way for a more personalized and effective learning experience. AI's prowess in processing vast datasets and discerning intricate patterns enables it to decode individual differences with unprecedented accuracy. Whether considering socio-economic backgrounds, cultural nuances, or personal preferences, AI algorithms dissect data to unveil nuanced characteristics. This intricate understan

6.2.1 UNDERSTANDING THE LANDSCAPE: AI LITERACY

In order to successfully incorporate artificial intelligence (AI) into education, it is crucial to have a thorough understanding of both its technical and human components. In this sense, AI literacy assumes a key role, helping institutions and educators to negotiate the complex landscape of cutting-edge technologies. Stakeholders may harness the potential of AI to tailor educational experiences to the diverse requirements of learners by comprehending the intricacies of this technology. AI literacy entails providing instructors and students with the necessary knowledge and abilities to comprehend and engage with AI technologies proficiently. This encompasses fundamental principles like machine learning, data analysis, and algorithmic thinking, along with the ethical and societal consequences of AI utilization. Incorporating AI literacy into the educational curriculum facilitates the process of unravelling the complexities of AI and ensuring its accessibility to all learners. It cultivates the development of critical thinking and problem-solving abilities, which are important in the contemporary technology-oriented society. Teachers can employ artificial intelligence tools to customize learning experiences, offering personalized feedback and materials that specifically target the unique needs of each student.

Moreover, AI literacy enables educators to embrace and modify AI-driven instructional tools. These technologies have the potential to optimize teaching methods, simplify administrative duties, and promote student involvement. Teachers who possess AI literacy can more effectively enable debates on the ethical utilization of AI and adequately educate students for a future in which AI plays a crucial role in diverse industries. Institutions can promote a comprehensive comprehension of AI to ensure that learners are not only users of technology but also engaged contributors in influencing its advancement and implementation. Adopting a proactive approach to AI literacy is essential for establishing an educational environment that is inclusive and fair, and that fully utilizes the capabilities of AI.

6.2.2 Assessment Precision: A Cornerstone of AI in Education

Artificial intelligence significantly enhances the process of evaluation, which is a fundamental aspect of education. Excellently designed assessments are crucial for evaluating student comprehension. Artificial intelligence (AI) enables assessments to achieve a level of precision that is unattainable through traditional methods. Through data analysis, artificial intelligence (AI) identifies individual learning patterns, strengths, and areas that need improvement. By possessing this detailed and precise understanding, teachers can adapt their techniques, ensuring a more targeted and effective instructional approach. Furthermore, exams powered by artificial intelligence can offer instant feedback, enabling educators to immediately modify their teaching tactics and students to promptly rectify their learning routes. This interactive process increases the acquisition of knowledge by promoting a constant cycle of enhancement and adjustment. Artificial intelligence (AI) has the ability to support several types of assessments, including multiple-choice examinations and intricate, open-ended questions, which allows for a thorough evaluation of students' abilities.

6.2.3 Learner–Instructor Interaction: The Human Touch in the Digital Realm

Through data analysis, artificial intelligence may discern individual learning habits, strengths, and areas that need improvement. Equipped with this highly refined knowledge, teachers can adapt their approach, guaranteeing a more targeted and effective educational method. Assessment is an essential element of education, and AI greatly enhances the accuracy of assessments compared to previous methods [4]. AI improves learner–instructor connection by facilitating customized communication. AI tools can be utilized by teachers to offer personalized advice and assistance, catering to the individual requirements of each pupil. This individualized attention facilitates the development of better teacher-student relationships, which are essential for fostering student motivation and engagement. Moreover, AI has the capability to aid in the management of administrative activities, so allowing teachers to allocate more time towards engaging in direct contact with their students.

6.2.4 Individual Differences in Reasoning

The recognition of cognitive differences among pupils has long been acknowledged in the field of educational research citearticle. Artificial intelligence (AI) explores the

AI and Optimization for Enhanced Public Services 111

distinct variations in reasoning, providing significant insights into the diverse cognitive capabilities exhibited by students. AI assists educators in tailoring teaching methods to suit various cognitive styles by understanding how students engage in problem-solving and critical thinking. Through the utilization of AI, educators can establish adaptable learning environments that effectively cater to the diverse cognitive requirements of pupils. This flexibility guarantees that every student, irrespective of their preferred learning method or speed, has the necessary assistance to achieve success. The capacity of AI to examine extensive quantities of educational data also aids in recognizing trends and patterns that can inform more comprehensive educational plans, ultimately resulting in a more inclusive and efficient educational system.

6.2.5 AI for Student Assessment: Beyond Conventional Approaches

In addition to elucidating the distinctions between AI-based and conventional assessment methods, Gonz´alez-Calatayud's [5] systematic review illuminates the paradigm-shifting capacity of AI in student evaluation. By doing so, it underscores the potential of AI to fundamentally alter the manner in which student performance is assessed. By virtue of their adaptability and responsiveness, AI-powered assessment tools guarantee that evaluations are in accordance with the unique learning trajectories of each student.

6.3 DECISION-MAKING IN WELFARE AND IMMIGRATION

The integration of Artificial Intelligence (AI) into welfare and immigration decision-making is causing a substantial paradigm shift. This technological transition has the capacity to enhance service delivery, optimize resource allocation, and promote efficiency in government operations. With the increasing prevalence of AI applications, especially in refugee status determination (RSD) procedures and immigration law applications, it is imperative to thoroughly examine these applications.

6.3.1 Optimizing Resource Allocation in Welfare

AI plays a crucial role in welfare decision-making by enhancing the allocation of resources. By analyzing vast datasets, AI algorithms can identify patterns and Artificial intelligence (AI) significantly influences welfare decision-making through its ability to optimize resource allocation. Through the analysis of extensive datasets, AI algorithms discern patterns and trends that empower governments to allocate resources more efficiently. This is especially relevant in welfare programmes, where precise targeting of assistance is critical to maximize impact. The capacity of AI to process and interpret intricate data sets facilitates a more nuanced comprehension of the requirements of individuals. trends, enabling governments to allocate resources more effectively. This is particularly pertinent in welfare programs where the precise targeting of support is essential for maximizing impact. AI's ability to process and interpret complex data sets allows for a more nuanced understanding of the needs of individuals and communities.

6.3.2 IMPROVING SERVICE DELIVERY IN WELFARE PROGRAMMES

The incorporation of AI into welfare decision-making encompasses service delivery in addition to resource allocation. An instance of this is ChatGPT, a sophisticated AI model that showcases the capacity for collaboration between AI and humans in formulating comprehensive and individualized responses [5]. Within the realm of welfare, this results in enhanced efficiency and customization of services for individuals. The mutually beneficial association between AI and human decision-makers has the potential to create a welfare system that is more responsive and centred around citizens.

6.3.3 AI IN IMMIGRATION DECISION-MAKING: BALANCING EFFICIENCY AND HUMAN RIGHT

The application of artificial intelligence in immigration raises important questions about the appropriate balance between efficiency and the protection of human rights. The use of AI in immigration decisions, as discussed in academic discussions, is multifaceted. On the one hand, AI can improve the speed and accuracy of processing immigration applications. On the other hand, concerns arise regarding the potential impact on human rights, particularly when decisions with significant consequences are delegated to automated systems.

6.3.4 HUMAN RIGHTS IMPLICATIONS IN AI-DRIVEN IMMIGRATION DECISION-MAKING

The use of AI in immigration decision-making has spurred discussions about the implications for human rights. The sheer magnitude of the potential impact, as CIGI [6] has shown, calls for a careful assessment of the ethical and legal aspects. The possibility of biased algorithms and the possibility of discriminatory outcomes raises questions about accountability and fairness. Finding the right balance between using AI to increase efficiency and respecting human rights in immigration decision-making is a difficult task that governments must undertake.

6.3.5 AI's POTENTIAL IN REPROGRAMMING ALGORITHMIC HUMANITARIANISM

The application of artificial intelligence (AI) is causing a paradigm shift in refugee status determination procedures. Scholars have advocated for the reprogramming of algorithmic humanitarianism, which aligns AI technologies with sustainable human rights governance [6]. The potential of AI to analyze large datasets could streamline RSD procedures, ensuring a more impartial and equitable assessment of refugee claims.

6.4 FRAUD DETECTION

The integration of artificial intelligence (AI) has caused a notable change in the way fraudulent operations are detected and prevented in several industries, thanks to the use of advanced tools and techniques. The identification of fraudulent activity has become an essential element in safeguarding operational and financial systems from harmful behaviours. AI integration in fraud detection utilizes sophisticated algorithms to analyze large volumes of data in real-time, detecting trends and anomalies that may suggest fraudulent activity. These systems has the ability to acquire knowledge and develop, hence improving their precision and efficiency as time progresses. By utilizing artificial intelligence, organizations can enhance their ability to identify instances of fraud with more efficiency.

Additionally, AI can enable organizations to anticipate and preemptively address suspected fraudulent behaviours, thereby adopting a proactive stance towards fraud prevention.

6.4.1 Anomaly Detection with AI

Anomaly detection is an artificial intelligence technique that effectively identifies suspicious patterns indicative of fraud. Leewayhertz highlights that anomaly detection is highly effective in identifying behaviour that deviates from the norm, making it an essential tool for fraud prevention (Wang, 2020). This method entails observing data for departures from established standards, allowing for the detection of uncommon actions that require additional examination. Anomaly detection systems driven by artificial intelligence have the capability to analyze data from diverse sources such as transaction records, user behaviour logs, and network activities in order to identify any abnormalities. These systems employ machine learning models to comprehend the definition of typical behaviour and subsequently identify any deviations for further examination. The capacity to adjust to novel patterns of deceptive conduct guarantees the continued efficacy of these systems, even in the face of evolving fraud strategies.

6.4.2 AI and Machine Learning Strategies

The use of machine learning enables the analysis of large datasets, ensuring a proactive approach to fraud prevention [7]. These technologies enhance precision and productivity by proposing and revising rules for detection and notifications. Machine learning models provide the capability to detect intricate patterns and connections within data that could potentially go unnoticed by conventional rule-based systems. Through the ongoing acquisition of new data, these models improve their ability to forecast, making them more skilled at detecting any fraudulent activities. AI and machine learning employ two main strategies: supervised learning and unsupervised learning. In supervised learning, models are trained using labelled datasets that contain known cases of fraud. On the other hand, unsupervised learning discovers anomalies without any prior knowledge of what is considered fraudulent. Furthermore, methods like reinforcement learning allow systems to acquire optimal detection tactics by repeatedly testing and adjusting their performance, resulting in continuous improvement over time.

6.4.3 Preventing Bank Fraud with AI

Mytnyk conducted an investigation on the application of artificial intelligence (AI) in the banking industry to mitigate fraudulent actions. Mytnyk's study highlights various effective approaches to combat bank fraud, such as the utilization of artificial intelligence (AI), consortium data, biometric data, and advanced standardization techniques [8]. Artificial intelligence (AI) systems utilized in the banking sector possess the capability to examine transaction patterns, identify atypical behaviours, and highlight possibly fraudulent transactions for additional scrutiny. Biometric data, such as fingerprints and facial recognition, enhances security by verifying that transactions are carried out by authorized individuals. The utilization of consortium data, which entails the exchange of information regarding fraudulent activities among many institutions, facilitates the development of more resilient models by offering a more extensive dataset for training purposes. Implementing high-tech standardization enables uniform adherence to security standards across all institutions, hence increasing the level of difficulty for fraudsters attempting to exploit system weaknesses.

6.4.4 Government Fraud Reduction with AI and Machine Learning

The Brookings Institution emphasizes the significance of data mining, machine learning, and artificial intelligence in mitigating occurrences of government fraud. These technologies enhance the government's capacity to deter or thwart fraudulent activities, showcasing the wide range of applications of artificial intelligence (AI) [9]. AI in the public sector has the capability to analyze extensive amounts of transaction data, audit logs, and other information in order to detect trends that suggest the presence of fraud. AI has the capability to identify irregularities in tax filings, social security claims, and procurement processes. This helps to minimize cases of fraud and guarantees that resources are distributed accurately. Machine learning models can be taught using past data to identify typical fraudulent patterns and anticipate potential future instances of fraud. This proactive strategy allows government agencies to take action before substantial losses occur, so improving the overall integrity and efficiency of public services.

6.4.5 AI and Machine Learning for Advanced Data Analytics

The research study titled "Utilizing Machine Learning and Artificial Intelligence for Fraud Prevention" explores the application of ML and AI in detecting and preventing fraud by employing sophisticated data analytics and identifying anomalous patterns [10]. Advanced data analytics utilize intricate algorithms to analyze intricate information and extract significant insights that might guide tactics for detecting fraud. Artificial intelligence (AI) and machine learning have the capability to detect outliers, which are data items that differ greatly from the average and are often an indicator of fraudulent activity. Organizations can perform focused investigations to verify the occurrence of fraud by directing their attention on certain abnormalities. In addition, AI has the capability to automate the examination of extensive datasets, hence decreasing the amount of time and effort needed for manual evaluations. This allows for faster responses to potential threats [10].

AI and Optimization for Enhanced Public Services

6.5 INFRASTRUCTURE PLANNING

Artificial intelligence (AI) technologies are being integrated into the infrastructure industry, which is causing traditional processes to undergo a revolution, resulting in increased efficiency and opening up new opportunities for sustainable development. AI has emerged as a transformative force in the planning and optimization of new infrastructure projects.

6.5.1 Workflow of AI-Optimized Public Services

The depicted workflow (Figure 6.1) illustrates an AI-based framework for practical schedule optimization. Integrating artificial intelligence, this system streamlines schedule planning and optimization processes. By leveraging advanced algorithms and data-driven insights, the framework enhances the efficiency and practicality of scheduling tasks. The diagram showcases a comprehensive approach to schedule optimization, marking a significant advancement in utilizing AI for practical and effective scheduling solutions.

6.5.2 Optimizing Design and Planning Processes

It is essential to make use of artificial intelligence in order to maximize the effectiveness of the design and planning phases of infrastructure projects. By utilizing machine learning algorithms,

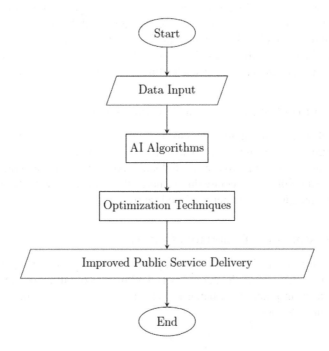

FIGURE 6.1 Workflow of AI-Optimized Public Services.

professionals are able to analyze large datasets, recognize patterns, and make decisions based on the data. This process speeds up the planning phase and ensures a more accurate and efficient outcome [11].

6.5.3 Enhancing Performance and Efficiency

The proactive decision-making that is enabled by the capability of artificial intelligence to analyze intricate data sets and deliver instantaneous insights is a means of reducing project setbacks and financial excesses [12]. Recent research, including that of Abdel-Kader, emphasizes the utilization of artificial intelligence (AI) methods in infrastructure projects as a means to enhance efficiency and performance.

6.5.4 Potential Applications in Construction

By utilizing machine learning and artificial intelligence, one is able to optimize resources, schedule projects, and mitigate unforeseen obstacles. These implementations contribute to the overall success of construction projects by streamlining procedures and improving results. The use of artificial intelligence in the construction industry encompasses a wide range of potential applications.

6.5.5 Unraveling AI Applications in Construction

Specifically, the capacity of artificial intelligence (AI) to automate mundane tasks, enhance precision, and facilitate predictive analytics emerges as a paradigm shift in the construction domain [13]. An exhaustive evaluation will be conducted by Abioye with the goal of elucidating the implementations of AI in the construction sector. The research will investigate the methodologies that are currently being used and will identify potential advantages and obstacles for AI applications in construction.

6.5.6 Impact on Civil Engineering and Infrastructure Development

The impact that artificial intelligence has had on civil engineering and the development of infrastructure is significant. The article that examines the role of AI in civil engineering places an emphasis on its influence on design and planning. AI tools enable the creation of infrastructure that is more efficient and sustainable, which is in line with the changing requirements of modern society.

6.5.7 Reshaping Civil Engineering Practices

Across the board, artificial intelligence is improving precision, optimizing workflows, and ultimately strengthening the durability and robustness of infrastructure. This shift in civil engineering practises is being brought to light by advancements in bridge design and construction.

6.6 CITIZEN INTERACTION

Artificial Intelligence (AI) is playing a crucial role in transforming citizen involvement and enhancing the efficiency of public services. The adoption of AI-powered tools to effectively address citizen inquiries represents a significant shift in how governments interact with their constituents. This comprehensive examination explores the primary uses of AI in government services, focusing on its ability to efficiently handle consumer inquiries and improve the overall quality of public service. AI technologies are revolutionizing both the citizen-government interaction and the effectiveness and openness of governmental operations. Through the incorporation of artificial intelligence (AI) into public service platforms, governments may guarantee that individuals are provided with prompt, precise, and tailored answers to their inquiries, thereby promoting a more involved and knowledgeable citizenry.

6.6.1 AI-Powered Chatbots: The Frontline of Citizen Engagement

AI chatbots are the main method for efficiently addressing citizen inquiries. These intelligent technologies utilize natural language processing to understand and provide immediate replies to public issues. AI-powered chatbots offer customized and efficient responses to residents, whether they require assistance, information on public services, or clarification on policy. This greatly enhances the overall experience of citizens. AI chatbots are capable of addressing a diverse array of inquiries, ranging from basic queries regarding government services to intricate requests necessitating comprehensive information. Through the automation of repetitive duties, these chatbots enable human agents to dedicate their attention to more intricate matters, consequently enhancing the overall effectiveness of government service provision. Furthermore, AI chatbots have the capability to function continuously throughout the day and night, guaranteeing that individuals have the ability to obtain support whenever needed.

6.6.2 Conversational AI in the Public Sector: A Driving Force for Success

Conversational AI, a cutting-edge kind of AI-driven technology, demonstrates significant potential in the public sector. These instruments, sometimes referred to as "advanced copilots," facilitate interactive and dynamic communication between citizens and government services. Conversational AI ensures that users receive accurate and personalized information by providing detailed responses and guiding them through complex procedures. This helps to create a positive view of public services. These sophisticated AI systems possess the capability to comprehend and analyze natural language, thereby enabling interactions that closely resemble those between humans. Furthermore, these systems have the capability to enhance their replies by leveraging past interactions, hence continuously enhancing their effectiveness in addressing the requirements of individuals. Conversational AI fosters trust and confidence in public services by offering precise and reliable information.

6.6.3 Practical Applications of AI in Government

An analysis of the practical applications of artificial intelligence (AI) in the public sector, as highlighted by V7 Labs, reveals the various ways in which AI enhances the efficiency of public services. AI serves as a versatile tool for governmental organizations by helping manage large amounts of public data and improving decision-making processes. Effectively addressing public inquiries is just one facet of the significant impact that AI has on enhancing government operations. AI is also applied in several domains, including predictive analytics for policy formulation, fraud identification, and resource distribution. For example, artificial intelligence (AI) has the capability to examine extensive datasets in order to detect and understand patterns and trends. This information may then be used to make informed policy decisions, ensuring that resources are distributed to the areas where they are most necessary. AI algorithms in fraud detection can discover anomalous patterns in data, which may suggest fraudulent actions, enabling prompt intervention.

6.6.4 Revolutionizing Citizen Services with AI

Implementing AI-driven technologies can optimize customer service by delivering prompt and precise responses to citizens. Not only does this enhance satisfaction, but it also contributes to the establishment of trust in the government's services [14]. AI-powered systems can potentially be used with other technologies like blockchain to improve transparency and security. AI has the capability to automate the process of verifying papers and transactions, hence minimizing the likelihood of errors and fraudulent activities. AI fosters a more robust connection between the government and its constituents by guaranteeing the delivery of dependable and uniform services to individuals.

6.6.5 The Government and Public Services AI Dossier

Public workers utilize artificial intelligence (AI) to detect fraudulent activities, strategize infrastructure projects, disburse welfare payments, process immigration decisions, and efficiently address citizen inquiries. By incorporating AI, the government becomes more responsive and focused on the needs of its citizens, as emphasized in a detailed AI report [15]. The report presents multiple instances where AI has been effectively utilized to enhance the delivery of public services. For instance, artificial intelligence (AI) has been employed to optimize the handling of welfare applications, guaranteeing the prompt and precise distribution of benefits. Artificial intelligence (AI) is utilized in immigration to effectively handle the substantial amount of data linked to visa applications, hence enhancing the efficiency and transparency of the process.

6.7 JUDICIAL PROCESSES

Artificial intelligence (AI) is revolutionizing the process of bail hearings by being integrated into judicial proceedings. AI is a valuable tool that aids judges in making educated decisions, enhancing the speed and fairness of the adjudication process.

AI and Optimization for Enhanced Public Services 119

The potential of AI goes beyond bail hearings and has an impact on many judicial processes, such as case management, legal research, and evidence review. The incorporation of artificial intelligence (AI) in the judicial system seeks to improve the efficiency, precision, and impartiality of legal procedures, guaranteeing that justice is delivered promptly and fairly.

6.7.1 AI as an Aid to Human Decision-Making

These technologies utilize machine learning algorithms to examine broad and relevant data, providing judges with extra insights that can inform their verdicts. Algorithmic tools have grown more used in court systems, specifically in bail hearing inquiries, to assist in decision-making in criminal cases [16]. These tools offer judges support in rendering rulings. AI can assist in forecasting case results using past data, recognizing pertinent precedents, and emphasizing crucial facts. AI enhances the quality of legal rulings by offering judges thorough and evidence-based insights, enabling them to make better-informed decisions.

6.7.2 Pretrial Risk Assessment Algorithms

The use of pretrial risk assessment algorithms is a significant implementation of artificial intelligence (AI) in the determination of bail hearings. These algorithms are crucial in assessing the possible risk that a person may pose if given bail. To generate risk scores, judges can evaluate many factors like the defendant's criminal record, likelihood of fleeing, and the seriousness of the alleged crime, together with other relevant data [17]. The purpose of these algorithms is to reduce bias and maintain uniformity in bail determinations. Nevertheless, it is imperative to consistently oversee and revise these systems in order to avert potential prejudices that may emerge from the data employed to train the models. Ensuring the transparent and accountable utilization of AI in risk assessments is crucial for upholding public confidence in the judicial system.

6.7.3 Capacity Building for Judicial Systems

Given the potential of artificial intelligence (AI) to influence judicial procedures, global initiatives are underway to enhance the capabilities of judicial systems. The objective of these endeavours is to guarantee that judges possess the ability to utilize AI as an additional tool while maintaining strict adherence to the essential principles of fairness and neutrality [18]. Training programmes and workshops focused on AI technologies and their applications in the court are crucial for providing judges and legal professionals with the requisite expertise and understanding. This capacity building guarantees the efficient integration of AI into the judicial process, amplifying its advantages while minimizing potential hazards.

6.7.4 Fairness and Efficiency Gains

Artificial intelligence enhances the fairness of the adjudication process by considering a broader set of elements and assessing evidence objectively, without any preconceived notions. These are only a handful of the notable advantages that accompany the utilization of AI in bail proceedings. AI systems can analyze information objectively, minimizing the impact of inadvertent biases that may affect human decision-makers. Furthermore, AI has the capability to optimize administrative chores, minimize the accumulation of pending cases, and expedite the resolution of legal matters. This increase in efficiency not only benefits the judicial system but also enhances access to justice for citizens, guaranteeing that legal procedures are conducted promptly and efficiently.

6.7.5 Ensuring Transparency and Accountability

Ensuring transparency and accountability is crucial as artificial intelligence (AI) gets more deeply incorporated into judicial operations. Judges must possess a thorough comprehension of the algorithms and procedures that form the foundation of the AI technologies used in bail proceedings. Having this information allows them to assess the results with careful judgement, retain control over the decision-making procedure, and uphold the principles of legal governance. It is crucial to establish unambiguous norms and ethical criteria for the utilization of artificial intelligence in the judicial system. Conducting regular audits and assessments of AI systems can assist in guaranteeing their dependability, impartiality, and responsibility, therefore protecting the integrity of judicial processes.

6.8 HEALTHCARE TRIAGE

The integration of intelligent optimization techniques into healthcare management is radically transforming how healthcare professionals prioritize, assess, and address patient needs. The healthcare business is being greatly influenced by artificial intelligence, especially in the important area of case triage, which focuses on guaranteeing quick and effective replies. AI technologies are augmenting the precision and swiftness of triage choices, better patient outcomes, and optimizing the utilization of healthcare resources. Through the utilization of artificial intelligence, healthcare practitioners have the ability to provide more individualized and prompt medical attention, ultimately revolutionizing the overall experience for patients.

6.8.1 Transforming Triage with AI

Triage, a vital operation in the healthcare sector, allocates precedence to patients based on the severity of their diseases. Artificial intelligence (AI) is transforming this process by utilizing advanced algorithms and machine learning techniques. These technologies analyze a wide range of patient data, including medical history, vital signs, and other relevant information, to accurately assess the urgency of each particular case [19]. AI-powered triage systems rapidly detect high-risk patients and provide priority to their treatment, guaranteeing that individuals in critical condition receive prompt care. The prompt and precise evaluation is especially crucial in urgent situations, where quick intervention can greatly influence patient results.

6.8.2 BENEFITS OF AI IN THE TRIAGE PROCESS

The application of artificial intelligence (AI) in the triage process offers notable benefits, such as enhanced efficiency, precision, and patient results. AI ensures that urgent cases are promptly addressed by efficiently processing and analyzing data, leading to improved allocation of resources and reduced reaction times in emergency situations. AI technologies can alleviate the workload of healthcare practitioners by automating repetitive evaluations and providing assistance in making decisions. This enables healthcare personnel to concentrate on more intricate cases and provide care of superior quality.

6.8.3 PREDICTIVE ANALYTICS FOR PROACTIVE HEALTHCARE

The intelligent optimization of artificial intelligence extends beyond triage to encompass predictive analytics. Machine learning algorithms possess the capability to identify trends and predict possible health issues before they escalate in severity. By adopting a proactive strategy, healthcare practitioners can effectively implement preventative measures, resulting in a decrease in the total strain on the healthcare system and an enhancement in patient outcomes. AI can facilitate early intervention and personalized care regimens by identifying individuals who are at risk of developing chronic illnesses or having problems. This proactive healthcare paradigm not only enhances patient well-being but also diminishes healthcare expenses by averting unnecessary hospitalizations and treatments.

6.8.4 ENHANCED HEALTHCARE MANAGEMENT

Artificial intelligence (AI) enhances hospital operations by analyzing extensive datasets, leading to greater efficiency, reduced costs, and enhanced patient satisfaction [20]. Intelligent optimization strategies in healthcare management extend beyond individual instances. Artificial intelligence (AI) has the ability to optimize administrative procedures, oversee inventory management, and forecast patient admission rates, so enhancing the overall efficiency and effectiveness of hospital operations. This holistic approach to healthcare administration improves the overall efficiency of healthcare provision.

6.8.5 ETHICAL CONSIDERATIONS AND CHALLENGES

While artificial intelligence (AI) provides significant benefits, it is crucial to recognize and address the ethical issues and challenges that come with its adoption. Responsible utilization of AI in the healthcare industry necessitates the protection of patient privacy, the reduction of algorithmic prejudices, and the establishment of transparent decision-making protocols. Moreover, ongoing education for healthcare professionals is crucial for maximizing the potential of new technologies while upholding ethical norms. It is essential to tackle these ethical concerns in order to uphold trust in AI-powered healthcare systems. By ensuring the ethical and responsible use of AI tools, we can fully harness their promise to enhance healthcare outcomes.

6.9 DRONE PATH PLANNING

Unmanned aerial vehicles (UAVs), popularly known as drones, are now essential in various industries, including delivery services and monitoring. In order to maximize their performance, it is crucial to assess the efficiency of drone paths. In this regard, Artificial Intelligence (AI) plays a crucial role by using advanced optimization algorithms to discover the most efficient routes. AI technologies augment the functionalities of drones, empowering them to traverse intricate surroundings, evade obstacles, and accomplish jobs with enhanced accuracy and efficiency. These technological developments are fueling innovation in other industries, such as logistics, agriculture, and public safety.

6.9.1 THE LANDSCAPE OF DRONE PATH PLANNING ALGORITHMS

A plethora of optimization algorithms have been developed in response to the quest for optimal drone routes. Researchers have explored methodologies that are influenced by biology (s23063051) and the human brain (10.3389/fnbot.2023.1111861). The main goal of these algorithms is to enhance the self-control of unmanned aerial vehicles, allowing them to navigate complex environments with agility and precision. Biologically inspired algorithms, such as genetic algorithms and ant colony optimization, imitate natural processes to address intricate path-planning problems. These algorithms are efficient at identifying the most optimal paths in ever-changing and uncertain surroundings.

6.9.2 BIO-INSPIRED OPTIMIZATION ALGORITHMS

Bio-inspired algorithms, such as Particle Swarm Optimization (PSO), have garnered significant interest due to their ability to generate independent flight paths for unmanned aerial vehicles (UAVs) [21]. The PSO algorithm outperforms other algorithms due to its ability to optimize the operational environment for drones, resulting in the facilitation of adaptive and efficient path planning. PSO algorithms mimic the collective behaviour of birds flocking or fish schooling in order to efficiently explore and exploit search areas. This methodology enables unmanned aerial vehicles to adjust to dynamic circumstances and identify optimal routes in the present moment.

6.10 PUBLIC SECTOR INTEGRATION

In recent times, there has been a notable increase in the implementation of Artificial Intelligence (AI) within public sector entities. This surge is transforming the governance domain by introducing fresh prospects for enhanced efficiency, decision-making, and citizen engagement. In this discourse, we shall explore the fundamental insights pertaining to the growing prevalence of AI in the public sector and the considerable advantages it imparts.

6.10.1 Adoption Trends

The implementation of artificial intelligence (AI) in public administration is a complex occurrence. A systematic literature review examines the various factors that influence this adoption, providing a nuanced perspective on AI diffusion and adoption [22]. Organizations are attempting to navigate this intricate terrain by taking into account technological, organizational, and environmental considerations. AI adoption is not a fixed procedure, but rather a dynamic one, with organizations consistently modifying their AI strategies to correspond with the ever-changing environment.

6.10.2 Benefits of AI in the Public Sector

For the purpose of optimizing policy design, decision-making processes, and citizen communication, governments are increasingly turning to artificial intelligence (AI). The potential benefits of AI in government extend beyond the enhancement of operational efficiency; it also enables the utilization of data-driven insights to generate novel forms of value creation that contribute to improved governance capabilities.

1. Enhanced Decision-Making: Artificial intelligence (AI) makes it easier for policymakers to make decisions based on data, providing them with insights that are both practical and implementable. This not only improves the quality and effectiveness of policy choices but also optimizes administrative procedures. With the assistance of predictive analytics and pattern recognition, governments are able to make decisions that are well-informed.
2. Improved Citizen Engagement: The implementation of artificial intelligence (AI) in public sector operations is fundamentally altering the way citizens interact with government. By utilizing chatbots and virtual assistants that are driven by AI, citizens are able to receive personalized and timely responses. This facilitates effortless access to information, support, and engagement with government services, thereby promoting a governance model that is more focused on the needs and interests of the populace.
3. Operational Efficiency: The automation capabilities of artificial intelligence play a significant role in improving operational efficiency. These capabilities automate routine and repetitive tasks, which enables employees in the public sector to concentrate on more complex and strategic aspects of their roles. This not only optimizes resource utilization but also reduces the risk of human errors.

6.11 IMPROVING PUBLIC ADMINISTRATION

Artificial intelligence (AI) is causing a shift in the way public administration operates by simplifying the process of formulating policies, enhancing the quality of services provided, and streamlining the management of internal operations. This emerging technology has the potential to completely transform the way government operates by making services more effective, data-driven, and centred on the requirements of citizens.

6.11.1 Transformative Implications for Policy-making

Artificial intelligence has had a significant impact on policy-making, ushering in a new era of evidence-based decision-making. Research highlights the need for more empirical studies that focus on the public sector in order to understand the process-wise implications of AI in public governance [23]. By leveraging AI, governments gain access to predictive analytics and pattern recognition, which empowers policymakers to make informed decisions.

1. Data-Driven Decision-Making: AI makes it possible to analyze large datasets, which gives policymakers access to insights that they might not have otherwise. This data-driven approach improves the precision and effectiveness of policy decisions.
2. Predictive Analytics: Policymakers can reap the benefits of great foresight by employing AI algorithms to forecast future trends and potential results. This proactive approach contributes to the formulation of policies that are more effective and preventative.

6.11.2 Optimizing Public Service Delivery

Chatbots, virtual assistants, and automation tools that are powered by artificial intelligence all contribute to a more responsive and streamlined public service experience. The adoption of AI in public service delivery is a paradigm change since it provides citizens with services that are both effective and individualized.

1. Citizen-Centric Engagement: Through the use of avatars and virtual assistants, artificial intelligence makes it possible to have individualized interactions with individuals, responding to their inquiries and providing support. This level of active participation increases overall satisfaction and confidence in public services.
2. Operational Efficiency: Automation of routine tasks through the use of artificial intelligence (AI) helps to maximize operational efficiency. This enables employees in the public sector to focus on more complicated responsibilities while simultaneously reducing the likelihood of human error and the amount of labour that is involved with repeated work.

6.11.3 Streamlining Internal Management Processes

Public administrations are currently facing a paradigm shift in the method in which they monitor staff, resources, and information. The incorporation of intelligent solutions enhances the efficiency of internal procedures, which is a positive development.

1. Resource Optimization: Artificial intelligence has the potential to improve the efficiency of public sector organizations by providing assistance in the

AI and Optimization for Enhanced Public Services

allocation and utilization of resources. This includes the design of infrastructure, the allocation of budgets, and the management of workforces.

2. Data Management: Artificial intelligence has the potential to improve the efficiency of public sector organizations by providing assistance in the allocation and utilization of resources. This includes the design of infrastructure, the allocation of budgets, and the management of workforces.

6.12 CONCLUSION

The amalgamation of optimization techniques and Artificial Intelligence (AI) has initiated a paradigm shift in the realm of public services, fundamentally reshaping efficiency and citizen engagement. This influence transcends multiple domains, including intelligent government systems and public transportation. The emergence of intelligent government, enabled by AI, not only augments the productivity of public personnel but also carries substantial ramifications for the future of governance. The overarching influence of artificial intelligence and optimization is encapsulated in significant improvements in public service delivery, internal management processes, and policy formulation, contributing to a more streamlined, data-driven, and citizen-centric public administration. Furthermore, scholarly reviews demonstrate the potential of AI's optimization and decision-support techniques in effective policymaking, data mining, and opinion analysis for greater efficiency. Figure 6.2 serves as a visual representation, showcasing the impact of diverse optimization techniques on public services. Ranging from "Genetic Algorithms" to "Algorithmic Efficiency Improvement," each technique is depicted on the vertical axis, with corresponding impact scores simulated on the horizontal axis. The horizontal bar chart allows for a clear comparison of impact scores, with 'skyblue' colouring enhancing visual appeal. This concise and informative visualization enables stakeholders to quickly grasp

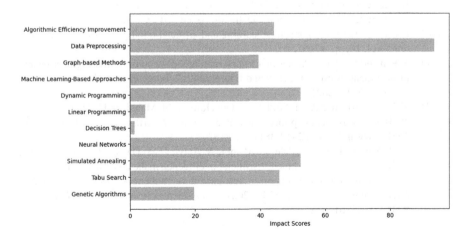

FIGURE 6.2 Impact of AI and Optimization Techniques on Public Services.

the relative contributions of different optimization methods to overall public service improvement.

REFERENCES

[1] S. P. Osborne, M. Cucciniello, G. Nasi, and E. Zhu, "Digital transformation, artificial intelligence and effective public services: Challenges and opportunities," *Global Public Policy and Governance*, vol. 2, no. 4, pp. 377–380, 2022.

[2] J. Reis, P. Esp´ırito Santo, and N. Melao, "Artificial intelligence in Government services: A systematic literature review," vol. 4, pp. 241–252, 2019.

[3] "The role of artificial intelligence in government — ey.com," www.ey.com/enes/government-public-sector/why-ai-and-the-public-sector-are-a-winning-formula, [Accessed 28-11-2023].

[4] K. Seo, J. Tang, I. Roll, S. Fels, and D. Yoon, "The impact of artificial intelligence on learner– instructor interaction in online learning," *International Journal of Educational Technology in Higher Education*, vol. 18, no. 1, pp. 1–23, 2021.

[5] Y. K. Dwivedi, N. Kshetri, L. Hughes, E. L. Slade, A. Jeyaraj, A. K. Kar, A. M. Baabdullah, A. Koohang, V. Raghavan, M. Ahuja *et al.*, ""so what if chatgpt wrote it?" multidisciplinary perspectives on opportunities, challenges and implications of generative conversational AI for research, practice and policy," *International Journal of Information Management*, vol. 71, p. 102642, 2023.

[6] N. Ahmad, "Refugees and algorithmic humanitarianism: Applying artificial intelligence to rsd procedures and immigration decisions and making global human rights obligations relevant to ai governance," *International Journal on Minority and Group Rights*, vol. 28, p. 12, 2020.

[7] B. Baesens, V. Van Vlasselaer, and W. Verbeke, *Fraud analytics using descriptive, predictive, and social network techniques: A guide to data science for fraud detection*. John Wiley & Sons, 2015.

[8] B. Mytnyk, O. Tkachyk, N. Shakhovska, S. Fedushko, and Y. Syerov, "Application of artificial intelligence for fraudulent banking operations recognition," *Big Data and Cognitive Computing*, vol. 7, no. 2, p. 93, 2023.

[9] D. M. West, "Using AI and machine learning to reduce government fraud — Brookings — brookings.edu," 2021. www.brookings.edu/articles/using-ai-and-machine-learning-to-reduce-government-fraud/, [Accessed 28-11-2023].

[10] P. Gupta, "Leveraging machine learning and artificial intelligence for fraud prevention," vol. 10, pp. 47–52, 2023.

[11] Y. Pan and L. Zhang, "Roles of artificial intelligence in construction engineering and management: A critical review and future trends," *Automation in Construction*, vol. 122, p. 103517, 2021.

[12] M. Y. Abdel-Kader, A. M. Ebid, K. C. Onyelowe, I. M. Mahdi, and I. Abdel-Rasheed, "(AI) in infrastructure projectsmdash;gap study," *Infrastructures*, vol. 7, no. 10, 2022. www.mdpi.com/2412-3811/7/10/137

[13] S. O. Abioye, L. O. Oyedele, L. Akanbi, A. Ajayi, J. M. Davila Delgado, M. Bilal, O. O. Akinade, and A. Ahmed, "Artificial intelligence in the construction industry: A review of present status, opportunities and future challenges," *Journal of Building Engineering*, vol. 44, p. 103299, 2021. www.sciencedirect.com/science/article/pii/S2352710221011578

[14] Z. Engin and P. Treleaven, "Algorithmic government: Automating public services and supporting civil servants in using data science technologies," *The Computer Journal*, vol. 62, no. 3, pp. 448–460, 2019.

[15] M. Choroszewicz and B. M̈aiḧaniemi, "Developing a digital welfare state: Data protection and the use of automated decision-making in the public sector across six eu countries," *Global Perspectives*, vol. 1, no. 1, p. 12910, 2020.

[16] C. Coglianese and L. M. B. Dor, "Ai in adjudication and administration," *Brook. L. Rev.*, vol. 86, p. 791, 2020.

[17] S. Greenstein, "Preserving the rule of law in the era of artificial intelligence (AI)," *Artificial Intelligence and Law*, vol. 30, no. 3, pp. 291–323, 2022.

[18] G. Canela, N. Burki, and S. Menon, "Unesco's judges' initiative: Training the custodians of the legal system on freedom of expression, access to information and the safety of journalists," *Max Planck Yearbook of United Nations Law Online*, vol. 25, no. 1, pp. 54–76, 2022.

[19] V. Boæifa´, "Using artifical intelligence in triage process: Benefits, challenges, and considerations," 2023.

[20] S. A. Alowais, S. S. Alghamdi, N. Alsuhebany, T. Alqahtani, A. I. Alshaya, S. N. Almohareb, A. Aldairem, M. Alrashed, K. Bin Saleh, H. A. Badreldin *et al.*, "Revolutionizing healthcare: the role of artificial intelligence in clinical practice," *BMC Medical Education*, vol. 23, no. 1, p. 689, 2023.

[21] S. Poudel, M. Y. Arafat, and S. Moh, "Bio-inspired optimization-based path planning algorithms in unmanned aerial vehicles: A survey," *Sensors*, vol. 23, no. 6, 2023. www.mdpi.com/1424-8220/23/6/3051

[22] R. Madan and M. Ashok, "Ai adoption and diffusion in public administration: A systematic literature review and future research agenda," *Government Information Quarterly*, vol. 40, no. 1, p. 101774, 2023. www.sciencedirect.com/science/article/pii/S0740624X22001101

[23] A. Zuiderwijk, Y.-C. Chen, and F. Salem, "Implications of the use of artificial intelligence in public governance: A systematic literature review and a research agenda," *Government Information Quarterly*, vol. 38, no. 3, p. 101577, 2021. www.sciencedirect.com/science/article/pii/S0740624X21000137

7 Architectural Pattern for Implementing XAI as a Service for Container-Orchestrated Machine Learning Model Deployments

Amit Chakraborty, Susmita Ganguly, and Saptarshi Das

7.1 INTRODUCTION

In the quickly changing field of artificial intelligence (AI) and machine learning (ML), deploying models that are not only powerful but also interpretable and transparent has become critical. Explainable AI (XAI) meets this demand by providing insights into how models make decisions, promoting confidence and accountability in AI systems. This paper presents an architectural approach for building XAI as a Service (XAIaaS) that is specifically designed for container-orchestrated environments, such as those controlled by Kubernetes. Using Kubernetes' scalability and flexibility, this design attempts to streamline the deployment, management, and explainability of ML models in a smooth and efficient manner. As AI systems become more integrated into crucial decision-making processes across multiple industries, the demand for transparency in AI-driven choices has increased. Regulatory frameworks, such as Europe's General Data Protection Regulation (GDPR), have stressed the importance of AI systems providing explanations for their outputs. Traditional AI deployment architectures frequently fall short of providing real-time explanations because of their complexity and additional computing overhead requirements. This gap necessitates a strong and scalable architecture solution that can incorporate explainability into the AI deployment process without sacrificing performance or scalability. Kubernetes, an open-source platform for automating the deployment, scaling, and operation of application containers has become the de facto container orchestration standard. This work contributes to the area by presenting a thorough architectural design that combines Kubernetes' capabilities with XAI's requirements.

Key contributions include:

Architecture Design: A detailed design blueprint that outlines the components and interactions needed to build XAIaaS in a Kubernetes ecosystem.

Case Study and Evaluation: An instructive case study showcasing the use of the proposed architecture in a real-world setting, as well as an assessment of its performance and scalability.

7.2 EXPLAINABLE ARTIFICIAL INTELLIGENCE (AI)

It is well proven that machine learning algorithms used as custom-built, or as-a-service endpoints are now not only a good-to-have feature for the business but are definitely a must-have for competitive advantage. Algorithms that can predict a customer risk profile aid decision-making for human consultants on the viability of justifying loan approval for an applicant. Similarly, algorithms can decide whether an invoice will be paid and if not what the trapped value is. For all these business processes that are differentiated and earn an edge over the competition by the usage of predictive analytics, explaining the output of the algorithm presents a problem. For business users and auditors alike, understanding why the algorithm predicts what it is predicting is crucial to answering and improving functional understanding [1]. There are different approaches to explainability, and interpretability is a subset of the whole [2].

7.3 INTERPRETABILITY

White box models can be interpreted. Examples include decision trees, their variants, logistic regression, etc. These are also called glass box models [1], [3]. The entropy equation of the decision tree renders itself interpretable by citing the point-wise probability for every observation:

$$\sum P(X) \log(X)$$

7.4 EXPLAINABILITY

For black box models such as Random Forest, KNN, Support Vectors or Neural network-based models, the interpretability becomes extremely challenging. In "MACHINE LEARNING: BETWEEN ACCURACY AND INTERPRETABILITY"

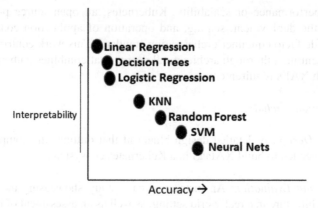

FIGURE 7.1 A non-quantitative representation of Interpretability vs. Accuracy.

paper [1] we see that the accuracy of a model increases at the cost of it being less interpretable. A graphical representation is as shown in Figure 7.1.

As researched in [2] to [4] we see that explainability can be arrived via:

Building interpretable models only like decision trees or logistic regression, which is clearly not domain justifiable or,
Drive explanations for complex models, called post-hoc [3], [4].

To achieve an explanation, further divisions can be made by considering:

Model Specific – explanations like Layer wise relevance propagation or integrated gradients for neural networks
Model Agnostic – Local Interpretable Model Agnostic Explanations (LIME) or Shapley Additives (SHAP).

The focus of this paper as described above is on an ambassador-based side car architecture for the deployment of explainable models, and hence, as a case study for the rest of this research work, we have focused on LIME.

7.4.1 Local Interpretable Model Agnostic Explanations (LIME)

LIME stands for Local Interpretable Model Agnostic Explanations [4], [5]. Its scope is local which means that it can be granulized to explain the model decision-making process at an observation level. Let's assume that on training a black box model we have come across a distribution that is highly non-linear and un-interpretable because it's not easy to explain why the model is making such decisions. LIME allows us to zoom into the individual observation we are interested in and build a simple model around it without worrying about the rest of the model. We can obtain an explanation for the observation of interest from that simple model which in turn is easily explainable [4].

7.4.2 IMPLEMENTATION OF LIME

To use LIME with any black box model for local interpretations the following approach was adopted [6]:

i. Random data points on the neighborhood of the observation of interest (X) were generated.
ii. The data point nearer to X will be assigned more weight than the point that's further away, the distance taken being the cosine distance [7], [8].
iii. The predictions of such points are made using the complex model 'f' and labels are generated.
iv. Now this data set with the labels is used to train 'g' and labels predicted.
v. The objective being to minimize the error f(z) and πx helps to add weights to those values of X that are nearer to the observation point of interest making LIME locally faithful [5], [8].

7.5 EXISTING ARCHITECTURE TO IMPLEMENT XAI USING CONTAINER ORCHESTRATION

The existing architecture coupled the implementation of the actual machine learning model with that of LIME. SHAP is equally applicable to this. Applications catering to an enterprise predictive analytics application typically consist of multiple machine learning models functioning to provide predictive endpoints to be consumed for downstream processing. These machine learning models are serialized for endpoint creation and are part of a machine learning operations workflow for feature selection, training, testing and validation. A very popular mechanism for the workflows is deploying Linux-based containers and this is managed [9], [10], operated and orchestrated through container orchestration platforms such as Kubernetes. Oftentimes these applications are coupled with mechanisms for determining an explanation via explainable AI mechanisms [11]. For locally interpretable and pointwise explanations LIME is one of the most commonly used algorithms. The current architecture described was implemented using container orchestration where an ingress controller handled the management of routing to the proper machine learning implementation deployment that runs the machine learning as well as the explainable pod [12]. Figure 7.2 gives a high-level overview of the existing architecture:

7.5.1 OBSERVATIONS FOR EXISTING ARCHITECTURE

The existing implementation was studied over a period of six months in their hypercare period through the incidents logged in the central ALM used by the software provider organization. Tickets that were considered for analysis were either marked as "Done" (successfully resolved) or "Closed" (Explanation provided and identified as a shortcoming of the system). A total of 326 tickets were analyzed and data on the root causes post-closure closure was taken. Table 7.1 gives the details of the same:

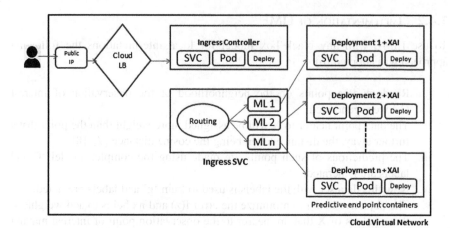

FIGURE 7.2 Existing architecture with ML model deployed along with XAI.

TABLE 7.1
Root cause vs number of tickets

Root Cause (RC)	Number of Tickets
RC1 – XAI Service not discoverable	121
RC2 – Inconsistent Remedial Policy	76
RC3 – Model confidence below threshold	48
RC4 – Service call failed	57
RC5 – Schema mismatch	16
RC6 – Outlier predicted	8

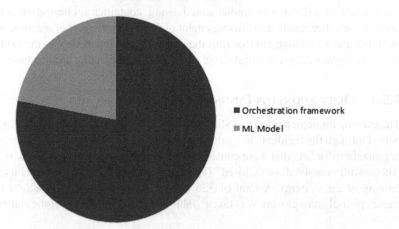

FIGURE 7.3 Distribution of tickets according to root causes and share between orchestration and ML modeling problems.

Architectural Pattern for XAI as a Service in ML Deployments

TABLE 7.2
Root cause vs cause of occurrence

Root Cause	Cause for Occurrence
RC1	Network challenges – All these endpoints connecting each other, other downstream applications or to client calls soon create a state of confusion that is not observable in terms of network traces, monitoring and efficient management.
RC2	Operational agility – Due to multiple endpoint monitoring, troubleshooting and securing operational agility becomes a challenge.
RC4	Application silos – Since different cloud platforms and libraries provide a wide variety of advantages over others the models use a heterogeneous mixture of such reusable components which creates application silos as they have varied runtimes and platforms.

A brief explanation of the root causes is given in Figure 7.3:

RC1 – Service exposed by the Kubernetes deployment specifically for XAI is not discoverable, that is the service endpoint is not available by any other pods that are calling the said service. This is due to the fact that the said service ingress cannot route the request to the corresponding pods.

RC2 – If a pod fails to respond the cluster policy should remedy the failure which is not happening. The remedy is generally to create another pod according to the deployment specification.

RC3 – The metrics returned by the model is not at par the set threshold. This causes the output from the end point to be unusable.

RC4 – The load balancer of the chosen cloud platform reaches to the ingress service but the call from ingress never happens which leads to the load balancer returning a null response (404) to the client.

RC5 – The input data that has the features to be predicted doesn't match with what the model is expecting.

RC6 – The predicted value is an outlier and is disregarded due to business logic.

RC1, RC2 and RC4 can contribute to problems for using a container orchestration framework and are the root cause for almost 77% of the closed incidents.

On subsequent discussions and architectural reviews RC1, RC2 and RC4 were summarized as shown in Table 7.2.

All these challenges stem from the fact that the logging, proxy management and configuration of these containers are taken care of by the main predictive container code base only. This added to the already heavy code base of model training, validation, best model selection and supporting continuous training of the models.

7.5.2 Telemetry of Existing Architecture

As per the scope of this research, only calls to XAI modules were recorded. The same can be easily obtained by adding a label in the response header and corresponding

TABLE 7.3
Metrics measurement

Metric	Definition	Measured for orchestrated platform with Kubernetes (AKS Ingress used)
Average response time per service endpoint	This is the response time averaged for requests for a span of 5 days (to have a uniformity in load)	~3.5 ms
Average response time for 10% fast query	For a valid input data for predicting what is the average response time for the first 10% of fastest Machine learning endpoints	~2.8 ms
Average response time for 10% slowest query	For a valid input data for predicting what is the average response time for the first 10% of slowest Machine learning endpoints	~4.7 ms
Failure rate per service	Number of times an ingress or egress call failed for a machine learning model endpoint per day	~12
Average Queue Depth	Number of parallel requests each machine learning model endpoint can handle on an average	~86

values aggregated [13], [14] for only the XAI calls. It is to be noted that we have considered only the metrics needed for architectural observability. The metrics and their average measures [3] calculated at a time period of 30 days are given in Table 7.3.

The above measures show that since the XAI API was deployed in the same pod, deployment and ultimately service of the ML model endpoint, the throttle problems augmented the failure rate, and service availability while utilization was very low (9%) [15], [17].

7.5.3 Proposed Architecture

We wanted to take the logging, configuration and proxies for the XAI part of the predictive containers separated out. These have to be uniform across the predictive containers and reusable to the extent that the set of language and frameworks can be reused across the predictive containers [16]. These nonfunctional components needed to be designed in such a way that they act in parallel to all the pods that are deployed in the orchestrated framework but can be owned by the same IaC or infrastructure team that supports the platform as a whole and need not be taken care of by data scientists who can concentrate upon the model training and validation [17]. Furthermore, these enablers need to also be deployed [18] as a feature alongside the main predictive model containers that must be co-located on the same host or

Architectural Pattern for XAI as a Service in ML Deployments

application. They should also be independently updated. Hence, we introduced the side car pattern for this container-orchestrated machine learning models deployed as microservices [19], [20].

For the said application we adopted this pattern in the orchestrated environment. We didn't replace the existing pattern as that would cause quite a considerable [21] code change but augmented the existing pattern [22]. Instead of the log, configuration, intra and inter networking and proxy being a part of the main machine learning container [23] we deployed it in a side container. The core functionality of the predictive application gets separated in this way from the side nonfunctional enablers as shown in Figure 7.4.

This will resolve the challenges described above. This enhances reusability also because once a side car for a predictive container is built it can be used with other predictive containers also keeping the organization standards and needs of nonfunctional requirements intact and repeatable [24]. This is a repeatable pattern and each of our predictive containers has a similar-in-nature side car alongside it.

Some considerations that came with the deployment are:

- Deployment was more tedious as side cars needed to be deployed with the predictive containers and this led to dependency injection in the deployment YAMLs [22], [25].
- Side car containers were developed using language and framework agnostic technology as that will allow using them in cross predictive analytics container applications.

The side cars' architectural design and implementation needed to be done as a separate library implementation which can be thought of as a separate daemon [20], [21].

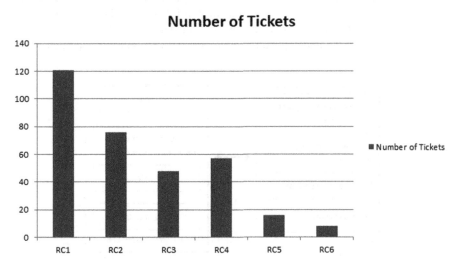

FIGURE 7.4 Core predictive containers deployed separately from the side car containers.

7.6 XAI APIS AS AMBASSADOR

The problem at hand was to create reusable templates for XAI APIs (LIME in this case). We used the ambassadors for the same where in the LIME implementation of any applicable machine learning containers were talking to the ambassador which in turn called the remote services of XAI for model agnostic explanations [26].

Figure 7.5 illustrates an architecture that integrates several moving parts such as multiple ML containers: ML1, ML2, MLn, and external services A, B, C with the use of the Ambassador layer.

XAI API endpoint deployed as a side car container – Side Car pattern is an architectural pattern that comes under the category of operational excellence and data management. Using this pattern, we separated out the XAI LIME part of the application into a separate process or container that works in parallel [27] to the actual machine learning endpoint containers to provide the needed isolation or encapsulation. The existing pods were just made lighter by removing the XAI component and creating separate containers for them. The implementation was done using Service Meshes. Service meshes can be thought of as a parallel runtime platform taking care of the security, networking and observability of container orchestration platforms. This is not a replacement for such orchestration platforms but is added as a layer that helps the individual components of such a platform to interact while all the telemetry, observability, service discovery and monitoring of these interactions are measured. Such a service mesh takes care of the inter service (read inter ingress service) communication, inter service security, and observability in terms of monitoring and tracing of distributed services and enhances their fault tolerance by automating circuit breaking and retires. The service mesh deployment essentially contains two planes: the control plane that contains the pods for the Kiali UI and Istio daemon and the data plane that contains the proxy containers that would run on each of the pods containing the predictive containers as the side car [28]. This is shown in Figure 7.6.

One way to inject the proxy side car containers is by using the Istio binaries which will add the YAML definitions in the main containers themselves. This actually impacts the IaC YAMLs and application code. So, we went with the approach of not

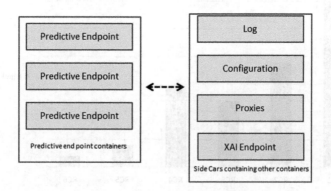

FIGURE 7.5 XAI APIs as Ambassadors of the ML containers.

Architectural Pattern for XAI as a Service in ML Deployments 137

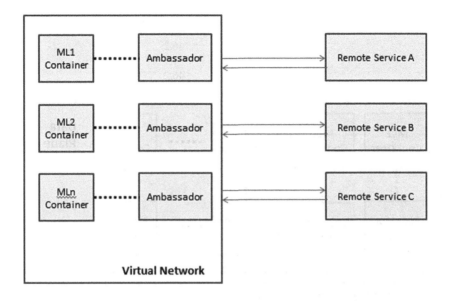

FIGURE 7.6 Core predictive containers deployed separately from the side car containers.

touching the existing YAMLs but adding a flag with Istio. This flag tells the Istio that we want to inject the proxy container in all of our pre-existing pods. This is mainly done by the pod called "side-car-injection yaml" [29]. This functionality is present in the Istiod daemon in the control plane [30]. This daemon monitors the pods from the data plane and whenever there is a pod being scheduled it injects the side car container or the proxy in the said pod [31]. This works on a namespace [32], [33] basis. The namespaces that are specially labeled will only be considered by Istio to be those in which to inject the containers. Once the pods are deployed we used Kiali to visualize the dependency map [34]. This sort of visualization inside the orchestrated framework helped in tracing defects and fixing connectivity issues. On further drilling, the problem with this interconnection can be summarized in Figure 7.7.

Improved Tracing with Jaegar – as part of the Istio set up used the tracing application Jaegar. As an example, for the interconnection service from the application the traces that we have obtained for the same [35].

As can be seen from the histogram (Figure 7.8) the response time for this service (as a node port) is quite high and sometimes of the order that's 100 times the average response time. This distributed tracing helped us to track the problem with the service more efficiently than what was deployed just as an orchestration framework without a service mesh [36]. What this means is that in the case of an external call also we have the ability to check the response time as the connection is now going through a service mesh proxy and hence return a HTTP 500 [37] error code if the response time > 1 sec. On introducing the response time constraint in the external visualization service call the response time came down to an average of 1.2 ms which is an 85% improvement. So what we observed is that without a service car

Code	Flags	% of Requests
200	-	46.2
503	UC	7.7
504	-	46.2

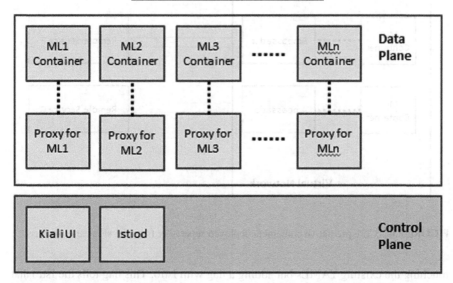

FIGURE 7.7 Distribution of error response code for the broken interconnection between A2C and PBI.

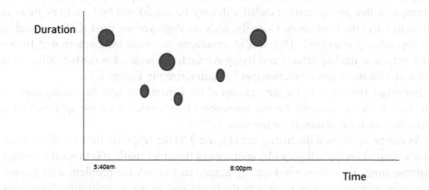

FIGURE 7.8 Histogram from Jaegar tracing for the service at fault.

proxy this problem would be rather untraceable and hence extremely difficult to solve [38]. Tickets corresponding to RC1 and RC2 showed a vast area of improvement along with telemetric improvements for this distributed tracing using service mesh with side car proxies.

Architectural Pattern for XAI as a Service in ML Deployments 139

7.7 IMPLEMENTATION VIA CONTAINER ORCHESTRATION

7.7.1 Service Discovery for the XAI Component

Service discovery in the proposed architecture was handled by ingress controllers. The Ingress controller used was the one provided by the public cloud platform, but the ingress service was designed to recognize the requests coming to the XAI [39] pod and direct them to the side car proxies.

```
apiVersion: v1
kind: Pod
metadata:
 name: xai-lime
spec:
 securityContext:
 runAsUser: 10
 runAsGroup: 20
 containers:
 - name: app
 image: xai/line:1.0
```

This definition creates an external address for the xai-lime service running as a proxy container to the service port of 7320. The definition above used annotations to specify more than one xai path if needed when multiple models of explanations must be used for the same machine learning model.

7.7.2 Process Containment for XAP API deployments through Orchestrated Containers

In this said architecture we used process containment to provide constraints in the usage of resources for processes running in the containers. This helps to limit the attack surface along with containing the processes. By having runtime process level security controls [41] and limiting to the principle of limiting privilege access we approached the problem of the ML containers to have to deal with access control [40], [42].

```
apiVersion: v1
kind: Pod
metadata:
 name: xai-lime
spec:
 securityContext:
 runAsUser: 10
 runAsGroup: 20
 containers:
 - name: app
 image: xai/line:1.0
```

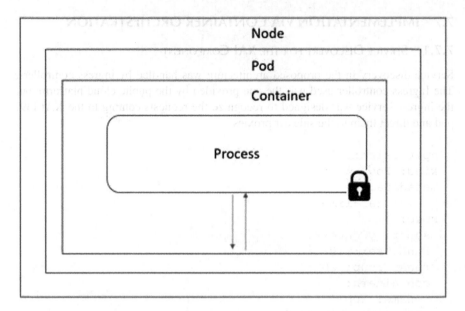

FIGURE 7.9 Inside the Kubernetes Nodes

Figure 7.9 shows the *securityContext* of the lime:1.0 image and specifies the *runAsUser* and *runAsGroup*. A generalized way for the user flag to be more non-restrictive is to set the *.spec.securityContext.runAsNonRoot* flag to true [42], [43].

```
securityContext:
    capabilities:
        drop: [ 'ALL' ]
        add: ['PROXY_SERVICE']
```

To avoid pod compromise, we have made the *securityContext* restrictive as shown above.

The containers that support XAI also take inputs from the relevant ML model but themselves cannot write to the container storage or blocks as they were designed to be ephemeral. We had set the *.spec.containers[].securityContext.readOnlyRootFile* to true so that any compromises don't enter the root file system or block. The method of securing containers at an individual level was made via higher-level constructs like Deployments and Cronjobs [44]. For a collection of pods to maintain a guardrail of security, we have used the Pod Security Standards (PSS) and Pod Security Admission (PSA) controller [43], [44]. This follows a policy-based approach from highly permissive to highly restrictive.

```
apiVersion: v1
kind: Namespace
```

Architectural Pattern for XAI as a Service in ML Deployments 141

```
metadata:
 name: xai-namespace
 labels:
 pod-security.kubernetes.io/enforce: baseline
 pod-security.kubernetes.io/enforce-version: v1
 pod-security.kubernetes.io/warn: restricted
 pod-security.kubernetes.io/warn-version: v1
```

Just like the above example, the XAI namespace was enforced with the baseline policy while warnings were raised for the restricted policy.

7.7.3 Network Segmentation for XAI Namespaces

Namespaces impose isolation constraints on containers by default belonging to different namespaces. In our architecture, we wanted to have a separate ambassador-based side car for our XAI-based containers and monitor secured inter-pod communication. We used network segmentation for the same [40], [44]. This helped us achieve multi-tenancy while maintaining application isolation. We used a combination of service ingress/egress along with the service meshes for filtering HTTP verbs and other L7 protocol parameters.

NetworkPolicy was used to define the rules for inbound and outbound networking connections for the pods. In our case we allowed inbound from ingress service, outbound to the specific ML container and back to an egress service for the final output.

```
kind: NetworkPolicy
apiVersion: networking.k8s.io/v1
metadata:
 name: allow-xai
spec:
 podSelector:
 matchLabels:
 app: xai-lime
 id: xai
 ingress:
 - from:
 - podSelector:
 matchLabels:
 app: ml-app1
 id: backend
```

In this example, we can see that the xai-lime pod can talk only to the ml-app1 pod so that the network segmentation is in place.

Securing the Configuration – The proposed architecture will have its primary uses during audits and business justification discussions regarding explainability [35], and will improve data and decision-making. So the architecture had to be designed in such a way that it does not live in isolation. Ordinarily, secrets in Kubernetes are not

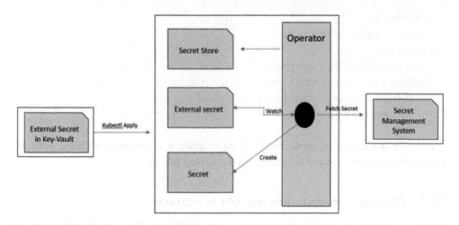

FIGURE 7.10 Secret Management System

encrypted but Base64 encoded. We integrated into our architecture Centralized Secret Management wherein we used the key-vault service to support an external secret management system.

Figure 7.10 explains how sidecar injection takes place in a Kubernetes context.

The following snippet shows how we used the *secretref* object and *secretaccesskeyref* inside the same to access the secret to enable the XAI pods to fetch the data from external PVC [21], [34], [40].

```
auth:
 secretRef:
 accessKeyIDSecretRef:
 name: cloudsm-secret
 key: access-key
 secretAccessKeySecretRef:
 name: cloudsm-secret
 key: secret-access-key
```

We used pod injection for secret management. Our proposed side car synced the secret from the cloud based external key value store to a local ephemeral volume that can be accessed by the XAI pods. This helps to update the secrets locally even when the secret management store rotates the secrets. Figure 7.11 shows the implementation of the same.

Use of GateKeeper for additional layer of security external to the cluster – To provide an extra layer of security and to assist in API throttling management we added a GateKeeper as an augmentation to our AbSC pattern. No direct calls to the XAI pods were allowed even when coming from a different VNET. For that we used APIM services so that the clients call the APIM service which will in turn pass the call to the ingress service with HTTPS header. To hide the XAI API call pattern we deleted the x-aspnet-version and x-powered-by headers and mask the URL Content property in

Architectural Pattern for XAI as a Service in ML Deployments

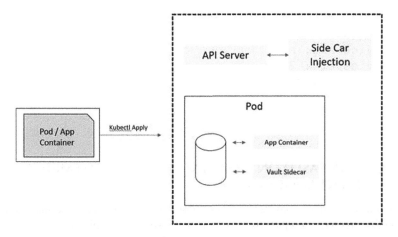

FIGURE 7.11 Side Car Injection

the requests. SSL offloading was handled at an application service level thus securing the ML model to XAI model calls also.

7.7.4 GATEWAY OFFLOADING

We added gateway offload to achieve operational excellence while maintaining a separation of the XAI pods as an ambassador inside car proxies. In here the main machine learning pods doesn't need to cater to the SSL certification management, Encryption, Token validation, Logging, Monitoring, or throttling [47]. These were added with the XAI pods and maintained as a side car to offload the gateway load. This resulted in lightweight ML pods and thus an enhanced modularity for our architecture. The calls to the XAI pods were managed through a queue-based load leveling design [48]. At the instance of the XAI API call, the call got registered in a message queue that sits inside the same VNET as the orchestration cluster and stores all messages. The message body consisted of the ML model and the data point of interest since this was implemented as an experiment primarily for a LIME explainability model. The client call, once through the ingress controller and service instead of going into the XAI pod directly, went to this message queue wherein it persisted. This controlled overloading of the XAI services while contributing a little to the latency [46], [49].

```
metadata:
 name: xai-lime
 annotations:
 nginx.ingress.kubernetes.io/fetch-explanations: /
spec:
 rules:
 - http:
 paths:
```

```
- path: /
  backend:
    serviceName: xai-lime
    servicePort: 7320
- path: /cluster-status
  backend:
    serviceName: cluster-status
    servicePort: 7320
```

7.8 RESULTS BASED ON OBSERVATIONS FROM TELEMETRY AND INCIDENT LOGGING

The application under consideration was observed through the same measurement system for telemetry and the observation noted is shown in Table 7.4.

While the average queue depth had increased the other telemetry metrics showed considerable improvement over a period of 6 months of observation. The incidents and their corresponding root causes showed the observations shown in Table 7.5.

While schema mismatches had increased due to the locally interpretable model not always being able to explain some predictions, other contributing causes had decreased.

TABLE 7.4
Metrics measurement

Metric	Measured for orchestrated platform with Kubernetes (AKS Ingress used)
Average response time per service endpoint	~ 2.8 ms
Average response time for 10% fast query	~1.2 ms
Average response time for 10% slowest query	~ 2.5 ms
Failure rate per service	~4
Average Queue Depth	~92

TABLE 7.5
Root cause vs number of tickets

Root Cause (RC)	Number of Tickets
RC1 – XAI Service not discoverable	18
RC2 – Inconsistent Remedial Policy	55
RC3 – Model confidence below threshold	42
RC4 – Service call failed	30
RC5 – Schema mismatch	28
RC6 – Outlier predicted	8

7.9 FUTURE WORK AND DIRECTIONS

While the proposed architecture for delivering explainable AI as a Service (XAIaaS) in Kubernetes-managed systems has shown potential, certain areas still require improvement to meet limitations discovered during implementation. One noteworthy problem identified was an increase in the frequency of schema incompatibilities when a generic side car approach was used inside the design. These incompatibilities were mostly caused by inconsistencies in data types and services, emphasizing the importance of more robust ways to address these concerns [45]. To address these problems, we suggest a number of architectural enhancements and expansions. Schema mismatches occur when expected and actual data formats or types differ, which can cause communication failures between services. In our architecture, these mismatches were compounded by the employment of generic side cars that failed to accommodate the unique data requirements of various services. To address this, we intend to incorporate more advanced data validation and transformation techniques into the side cars. This will entail developing flexible side cars that can dynamically adjust to the data schemas of the services they interface with. In addition, we want to use circuit breaking and fault injection mechanisms to improve the system's resilience. Circuit breaking enables the design to discover and isolate problematic services or components, avoiding cascading failures caused by schema incompatibilities. Fault injection, on the other hand, allows us to verify the system's robustness by intentionally introducing mistakes and seeing how well it recovers. These tactics will be critical in identifying. Addressing potential sites of failure will ensure that the design can handle schema mismatches more sympathetically. Another innovation we intend to provide is the incorporation of stateful patterns into service discovery. The existing architecture's stateless components can sometimes lead to inefficiencies, particularly when dealing with sophisticated, state-dependent processes. By adding stateful patterns, we hope to increase the efficiency and dependability of these operations. This will entail using Kubernetes' stateful sets and persistent volumes to keep the state across several instances and sessions.

This upgrade will rely heavily on service discovery. By including powerful service discovery techniques, the architecture can dynamically locate and engage with the right services, even as they scale or change over time. This will reduce. The likelihood of schema mismatches is reduced by ensuring that services always interface with the proper counterparts, hence maintaining system consistency. Currently, the suggested architecture does not fully enable the description of neural network models, which are becoming more common in modern AI applications. To solve this gap, we intend to incorporate neural network explanation capabilities into the design. This will need combining specialized tools and libraries, such as LIME (Local Interpretable Model Agnostic Explanations) and SHAP (SHapley Additive Explanations), which are intended to provide interpretable explanations for neural network outputs.

These tools will be integrated into the side car pattern, allowing them to coexist with the neural network models running in the Kubernetes cluster. This integration requires careful consideration of the computational cost and resource allocation. to ensure that the addition of explainability characteristics has no substantial impact on the performance of neural network models. The above-mentioned modifications aim to address the proposed XAIaaS architecture's present shortcomings by enhancing its robustness to schema mismatches, combining stateful patterns with service discovery, and providing neural network explanation support. These enhancements will not only strengthen and scale the architecture but will also make it more adaptable to a broader range of AI models and application cases.

By incorporating circuit breaking and fault injection, we improve the architecture's ability to manage faults and recover from failures, assuring continuous and dependable operation. The combination of stateful patterns and enhanced service discovery algorithms will increase the efficiency and consistency of state-dependent processes. Finally, incorporating neural network explanations ensures that the architecture can fulfill the growing demand for interpretable AI across a wide range of application domains.

These future development directions are critical for developing the suggested architecture into a more comprehensive and adaptable solution for deploying explainable AI in containerized systems. As we continue to enhance and expand the architecture, we expect it to serve as a solid platform for the deployment of transparent and trustworthy AI systems across a variety of industry sectors.

REFERENCES

1. Doran, Derek, Sarah Schulz, and Tarek R. Besold. "What does explainable AI really mean? A new conceptualization of perspectives." arXiv preprint arXiv:1710.00794 (2017).
2. Yang, Guang, Qinghao Ye, and Jun Xia. "Unbox the black-box for the medical explainable AI via multi-modal and multi-centre data fusion: A mini-review, two showcases and beyond." *Information Fusion* 77 (2022): 29–52.
3. Brik, Bouziane, et al. "A survey on explainable AI for 6G O-RAN: Architecture, use cases, challenges and research directions." arXiv preprint arXiv:2307.00319 (2023).
4. Kumara, Indika, et al. "FOCloud: Feature model guided performance prediction and explanation for deployment configurable cloud applications." *IEEE Transactions on Services Computing* 16.1 (2022): 302–314.
5. Adak, Anirban, et al. "Unboxing deep learning model of food delivery service reviews using explainable artificial intelligence (XAI) technique." *Foods* 11.14 (2022): 2019.
6. Zafar, Muhammad Rehman, and Naimul Mefraz Khan. "DLIME: A deterministic local interpretable model-agnostic explanations approach for computer-aided diagnosis systems." arXiv preprint arXiv:1906.10263 (2019).
7. Zafar, Muhammad Rehman, and Naimul Khan. "Deterministic local interpretable model-agnostic explanations for stable explain ability." *Machine Learning and Knowledge Extraction* 3.3 (2021): 525–541.
8. Mekki, Mohamed, et al. "XAI-Enabled fine granular vertical resources autoscaler." *2023 IEEE 9th International Conference on Network Softwarization (NetSoft)*. IEEE, 2023.
9. Zhong, Zhiheng, et al. "Machine learning-based orchestration of containers: A taxonomy and future directions." *ACM Computing Surveys (CSUR)* 54.10s (2022): 1–35.
10. Goebel, Randy, et al. "Explainable AI: The new 42." *Machine Learning and Knowledge Extraction: Second IFIP TC 5, TC 8/WG 8.4, 8.9, TC 12/WG 12.9 International Cross-Domain Conference, CD-MAKE 2018*, Hamburg, Germany, August 27–30, 2018, Proceedings 2. Springer International Publishing, 2018.

11. Saraswat, Deepti, et al. "Explainable AI for healthcare 5.0: Opportunities and challenges." *IEEE Access* 10 (2022): 84486–84517.
12. Angelov, Plamen P., et al. "Explainable artificial intelligence: An analytical review." *Wiley Interdisciplinary Reviews: Data Mining and Knowledge Discovery* 11.5 (2021): e1424.
13. Das, Arun, and Paul Rad. "Opportunities and challenges in explainable artificial intelligence (xai): A survey." arXiv preprint arXiv:2006.11371 (2020).
14. Core, Mark G., et al. "Building explainable artificial intelligence systems." AAAI (2006).
15. Rjoub, Gaith, et al. "A survey on explainable artificial intelligence for cybersecurity." *IEEE Transactions on Network and Service Management* 20.4 (2023): 5115–5140.
16. Eriksson, Håkon Svee. A user-centric approach to explainable AI in a security operation center environment. MS thesis. 2022.
17. Panagoulias, Dimitrios P., Maria Virvou, and George A. Tsihrintzis. "A microservices-based iterative development approach for usable, reliable and explainable AI-infused medical applications using RUP." *2022 IEEE 34th International Conference on Tools with Artificial Intelligence (ICTAI)*. IEEE, 2022.
18. Neghawi, Elie, et al. "Linking team-level and organization-level governance in machine learning operations through explainable AI and responsible AI connector." *2023 IEEE 47th Annual Computers, Software, and Applications Conference (COMPSAC)*. IEEE, 2023.
19. Arvind, C. S., et al. "Deep learning based plant disease classification with explainable AI and mitigation recommendation." *2021 IEEE Symposium Series on Computational Intelligence (SSCI)*. IEEE, 2021.
20. Sahu, Prateek, et al. "Sidecars on the central lane: Impact of network proxies on microservices." arXiv preprint arXiv:2306.15792 (2023).
21. Dattatreya Nadig, Nikhil. "Testing resilience of envoy service proxy with microservices." (2019). www.diva-portal.org/smash/get/diva2:1372122/FULLTEXT01.pdf
22. Pihlava, Ville. "Comparing service mesh proxy solutions." (2023). https://aaltodoc.aalto.fi/items/dcad9242-dedf-44d9-9039-c8da71a84a6b
23. Moens, Pieter, et al. "Edge anomaly detection framework for AIOps in Cloud and IoT." In *CLOSER*. 2023. www.scitepress.org/Papers/2023/118386/118386.pdf
24. Tkachuk, Roman-Valentyn, Dragos Ilie, and Kurt Tutschku. "Towards a secure proxy-based architecture for collaborative AI engineering." *2020 Eighth International Symposium on Computing and Networking Workshops (CANDARW)*. IEEE, 2020.
25. Li, Yana, Zepeng Zhang, and Lili Xu. "Multi-objective structural optimization of the aluminum alloy subway car body based on an approximate proxy model." *Advances in Mechanical Engineering* 14.5 (2022): 16878132221098898.
26. Rossi, Fabiana, Matteo Nardelli, and Valeria Cardellini. "Horizontal and vertical scaling of container-based applications using reinforcement learning." *2019 IEEE 12th International Conference on Cloud Computing (CLOUD)*. IEEE, 2019.
27. Rovnyagin, Mikhail M., et al. "ML-based heterogeneous container orchestration architecture." *2020 IEEE Conference of Russian Young Researchers in Electrical and Electronic Engineering (EIConRus)*. IEEE, 2020.
28. Naydenov, Nikolas, and Stela Ruseva. "Combining container orchestration and machine learning in the cloud: A systematic mapping study." *2022 21st International Symposium INFOTEH-JAHORINA (INFOTEH)*. IEEE, 2022.

29. Rovnyagin, Mikhail M., et al. "Algorithm of ML-based re-scheduler for container orchestration system." *2021 IEEE Conference of Russian Young Researchers in Electrical and Electronic Engineering (ElConRus)*. IEEE, 2021.
30. Toka, László, et al. "Machine learning-based scaling management for kubernetes edge clusters." *IEEE Transactions on Network and Service Management* 18.1 (2021): 958–972.
31. Ramos, Felipe, et al. "A machine learning model for detection of docker-based APP overbooking on kubernetes." *ICC 2021-IEEE International Conference on Communications*. IEEE, 2021.
32. Harichane, Ishak, Sid Ahmed Makhlouf, and Ghalem Belalem. "A proposal of kubernetes scheduler using machine-learning on cpu/gpu cluster." *Computer Science On-line Conference*. Springer International Publishing, 2020.
33. Wan, Ziyu, et al. "Kfiml: Kubernetes-based fog computing iot platform for online machine learning." *IEEE Internet of Things Journal* 9.19 (2022): 19463–19476.
34. Lee, Chun-Hsiang, et al. "Multi-tenant machine learning platform based on kubernetes." *Proceedings of the 2020 6th International Conference on Computing and Artificial Intelligence*. 2020.
35. Kanso, Ali, et al. "Designing a Kubernetes Operator for Machine Learning Applications." *Proceedings of the Seventh International Workshop on Container Technologies and Container Clouds*. 2021.
36. Chakraborty, Amit, Ankit Kumar Shaw, and Sucharita Samanta. "On a reference architecture to build deep-Q learning-based intelligent IoT edge solutions." In Rajdeep Chakraborty, Anupam Ghosh, Jyotsna Kumar Mandal, S. Balamurugan (Eds.) *Convergence of Deep Learning in Cyber-IoT Systems and Security*. Wiley, (2022): 123–146. https://onlinelibrary.wiley.com/doi/abs/10.1002/9781119857686.ch6
37. Singh, Pramod, and Pramod Singh. "Machine learning deployment using kubernetes." In Acquisitions Editor: Celestin Suresh John Development Editor: Laura Berendson Coordinating Editor: Aditee Mirashi (Eds.) *Deploy Machine Learning Models to Production: With Flask, Streamlit, Docker, and Kubernetes on Google Cloud Platform*. Apress Berkeley, CA, (2021): 127–146. https://link.springer.com/book/10.1007/978-1-4842-6546-8
38. George, Johnu, and Amit Saha. "End-to-end machine learning using kubeflow." *5th Joint International Conference on Data Science & Management of Data (9th ACM IKDD CODS and 27th COMAD)*. 2022.
39. Shim, Simon, et al. "Predictive auto-scaler for kubernetes cloud." *2023 IEEE International Systems Conference (SysCon)*. IEEE, 2023.
40. Kang, Peng, and Palden Lama. "Robust resource scaling of containerized microservices with probabilistic machine learning." *2020 IEEE/ACM 13th International Conference on Utility and Cloud Computing (UCC)*. IEEE, 2020.
41. Abdullah, Muhammad, Waheed Iqbal, and Abdelkarim Erradi. "Unsupervised learning approach for web application auto-decomposition into microservices." *Journal of Systems and Software* 151 (2019): 243–257.
42. Hamzehloui, Mohammad Sadegh, Shamsul Sahibuddin, and Ardavan Ashabi. "A study on the most prominent areas of research in microservices." *International Journal of Machine Learning and Computing* 9.2 (2019): 242–247.
43. Hamilton, Mark, et al. "Large-scale intelligent microservices." *2020 IEEE International Conference on Big Data (Big Data)*. IEEE, 2020.
44. Hilali, Anouar, Hatim Hafiddi, and Zineb El Akkaoui. "Microservices adaptation using machine learning: A systematic mapping study." *ICSOFT* (2021): 521–532.

45. Shaw, Ankit Kumar, et al. "Scalable IoT solution using cloud services–An automobile industry use case." *2020 Fourth International Conference on I-SMAC (IoT in Social, Mobile, Analytics and Cloud) (I-SMAC)*. IEEE, 2020.
46. Mohapatra, Debaniranjan, Amit Chakraborty, and Ankit Kumar Shaw. "Exploring novel techniques to detect aberration from metal surfaces in automobile industries." *Proceedings of International Conference on Communication, Circuits, and Systems: IC3S 2020*. Springer Singapore, 2021.
47. Chang, Chin-Chen, et al. "Signature gateway: Offloading signature generation to IoT gateway accelerated by GPU." *IEEE Internet of Things Journal* 6.3 (2018): 4448–4461.
48. Singh, Suneet Kumar, et al. "Offloading virtual evolved packet gateway user plane functions to a programmable ASIC." *Proceedings of the 1st ACM CoNEXT Workshop on Emerging in-Network Computing Paradigms*. 2019.
49. Yan, Peizhi, and Salimur Choudhury. "Deep Q-learning enabled joint optimization of mobile edge computing multi-level task offloading." *Computer Communications* 180 (2021): 271–283.

8 Quantum Machine Learning (QML) Algorithms for Smart Biomedical Applications

Inzimam Ul Hassan, Swati and Aakansha Khanna

8.1 INTRODUCTION

Quantum Machine Learning (QML) is an emerging field that integrates machine learning and quantum computing to develop novel algorithms with diverse applications, particularly in the biomedical domain [1].

In machine learning, data is analyzed, patterns are found, and predictions are made using mathematical and statistical models. Using special insights to expedite computation and provide a much more effective handling of massive volumes of data is the aim of QML.

In essence, QML algorithms work by first encoding data into quantum states, then manipulating those states using quantum operations, and then measuring the resultant states to extract data-related information [2]. Quantum computing uses qubits, which have the ability to be entangled with other qubits and lie in states, such as both 0 and 1, in contrast to classical computing, which uses classical bits (either 0 or 1). The invention of novel algorithms that are not possible to execute on classical computers is made possible by these special quantum features, which also allow for more effective processing of massive volumes of data. In some respects, QML is different from traditional machine learning. Firstly, QML algorithms may theoretically do calculations tenfold faster than classical algorithms since they work on quantum states [3-4]. Second, QML algorithms are typically created to make use of special facts of quantum computing, for carrying out tasks. Lastly, in order to execute calculations, QML algorithms need specific hardware, either a quantum computer or a quantum simulator. There are several possible advantages of QML for applications in the biomedical field. Large volumes of complicated data, such as genetic, clinical, and imaging data, are produced by the biomedical industry [5]. This data may be processed and analyzed more quickly and effectively with QML algorithms, which may result in more precise diagnosis, individualized treatment programmes, and successful drug discovery. QML algorithms, for instance, may be used to assess medical pictures for determining disease subtypes to forecast the course of a disease. The quantum support vector machine (QSVM) is a highly promising quantum machine learning (QML) algorithm specifically designed for biological applications. QSVM is the quantum equivalent of the support vector machine (SVM) technology, commonly employed in machine learning for classification tasks. The primary advantage of QSVM is its ability to leverage quantum computing to classify large and complex information efficiently. Figure 8.1 illustrates the key features and applications of QSVM in biological research. Numerous biological applications, such as illness detection, drug development, and genomic research, have already made use of QSVM.

Quantum variational autoencoders (QVAEs) are another intriguing quantum machine learning approach for biological applications [6-7]. Generative models, such as QVAEs, are capable of learning to encode and decode data, including genetic or medical pictures. By creating artificial data with QVAEs, machine learning models may be trained, increasing the precision of illness diagnosis and prognosis. For use in biomedical applications, more QML algorithms are being developed and tested in addition to QSVM and QVAEs. Quantum clustering methods, for instance, may be used to determine disease classes from genomic or medical picture information [8]. Treatment programmes and medication development initiatives can be improved by using quantum optimization techniques. Although QML has many potential uses in biomedical applications, there are also many obstacles to be solved. Furthermore,

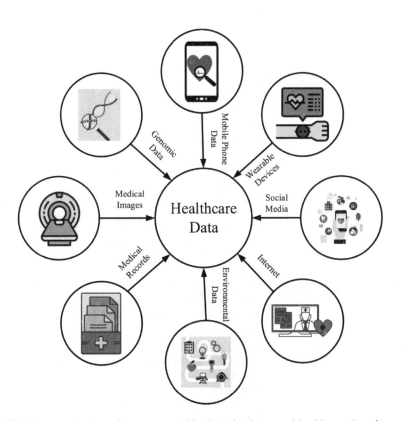

FIGURE 8.1 Application of quantum machine learning in several healthcare domains.

applying QML algorithms to real-world data can be challenging and sophisticated, requiring specific knowledge and software tools.

In conclusion, QML has the ability to completely transform the biomedical industry by making complicated data processing and analysis more effective [9-11].

8.2 REVIEW OF EXISTING LITERATURE

TABLE 8.1
Articles on QML

Year	Authors	Methods	Key Findings
2021	A Dabba et al.	The pre-processing techniques used for binary and multiclassification of gene data encompassed feature selection, cross-validation, support vector machines (SVM), and moth flame optimization. They also utilized the quantum moth flame optimization technique.	Reached the Colon 99% and Leukemial 97.2%. Ovarian cancer 99.5% Prostate cancer 96.2% CNS 91.1% 76% of the breasts 95% of the tumours 95.8% of 11 tumours 95.7% of brain tumours.
2021	G Sergioli et al.	Heistrom Classifier, a machine learning model inspired by quantum mechanics, uses a QML model in conjunction with pre-processing methods including feature extraction, normalization, and scaling to hide the background segmentation of the colonies.	
2021	Richard et al.	Pre-processing techniques such as data normalization, random cutting, PCA for features, and the D'Wave Quantum Ising model were employed.	Accuracy on six cancer cases was 91.22%.
2021	K Sengupta et al.	They deployed the QCNN model and used the state preparation and normalization technique for pre-processing.	95.57% accuracy was attained for COVID-19 data.
2020	A.T Jamal et al.	The study employed quantum genetic and quantum support vector machine techniques in conjunction with entropy-based multilevel thresholding and edge detection applications.	Image Detection Size: 150 x 150 pixels.

TABLE 8.1 (Continued)
Articles on QML

Year	Authors	Methods	Key Findings
2020	E. El-Shafeiy et al.	The QML strategy involves utilizing a QNN model together with pre-processing techniques like the Quick Reduction Feature Selection method. Additionally, the QNN model is trained using varying amounts of nodes.	With 30 nodes, the best outcomes were obtained. Accuracy of the QNN model was 92.334%.
2020	J Amin et al.	QCNN on the COVID 19 pictures using the QML approach.	Accuracy achieved 96%.
2020	A Seth et al.	For the ECG signal data, they employed various techniques.	An accuracy of 55% was achieved with QCNN, 53.55% with VQC, and 60% with Qboost.
2021	D Maheshwari et al.	Eight unique features were collected from each dataset using feature selection as part of the pre-processing step. Data normalization was then carried out using the Scalar and Min-Max approaches. Two encoding methods were used for QSVM and VOC models in order to prepare the state.	For the diabetes dataset, accuracy was attained using QSVM = 74.5%, VQC = 69%, and AEVQC = 74.4%.
2021	H Gupta.	They introduced shuffle sampling as a pre-processing method to the VQC model and used Feature Selection and Exploratory Data Analysis (EDA) as well.	74% accuracy attained at maximum for the diabetes dataset.
2021	Ishwarya M.S et al.	During the pre-processing phase of the quantum-inspired approach, they employed multi-attribute and multi-agent decision-making in combination with the ensemble technique.	90.5% was the maximum accuracy attained for the diabetes dataset.
2021	P. K Guru Diderot et al.	During the pre-processing phase, wavelet kernels were optimized and features were extracted from cancer mammography images using the HOP-WKELM model, which uses the Segment approach.	98.8% maximum accuracy for cancer data.

(continued)

TABLE 8.1 (Continued)
Articles on QML

Year	Authors	Methods	Key Findings
2021	D Pomarico et al.	Pre-processing involved the application of feature selection and cross-validation techniques on the QSVM model.	65.8% accuracy for data on breast cancer was attained.
2020	S. Saini et al.	Quantum support vector machines and VQC are employed by QML algorithms to treat breast cancer.	85% accuracy was attained using both QSVM and VQC.
2020	S. Chakraborty et al.	Hybrid quantum feature selection algorithm (HQFSA) is one example of a QML approach employed on a breast cancer dataset.	95% accuracy was attained for data on breast cancer.
2020	D. Sierra-Sosa et al.	The features were scaled and selected using techniques such as the ellipsoidal coordinate map, Stokes parameters, zero standard deviation normalization, ellipsoidal transform, and Poincare sphere. Additionally, the VQC model was utilized during the procedure..	The diabetes mellitus dataset has two and three characteristics. Using the Poincare sphere, they were able to attain the 72%.
2020	S Jain et al.	The QML quantum Boltzmann machine model was trained on the pre-processed lung cancer dataset through the use of feature selection, partitioning, normalization, and cross-validation procedures.	Achieved accuracy of 95.24%.
2020	D. Maheshwari et al.	By combining the voting model with the Qboost Quantum Ising model, the prediction of diabetic illness was enhanced.	Attained a 68.73% accuracy rate.
2020	V Iyer et al.	Images of skin cancer were downsampled using RBG colour for pre-processing, and the VQC model was autoencoded.	Achieved a maximum accuracy of 60%.
2020	H Yano et al.	State preparation, SDG, and Discrete feature mapping techniques— which were applied to the VQC model and utilized for QRAC— were all part of the pre-processing strategy.	VQC = 66.1% accuracy was attained, and VQC + QRAC = 72.6% accuracy.

TABLE 8.1 (Continued)
Articles on QML

Year	Authors	Methods	Key Findings
2020	A Bisarya et al.	The QML QCNN model was used to process the cancer data, and feature block, morphological pattern block, and state preparation approaches were used for classification.	Achieved a maximum accuracy of 98.1%.
2019	A Sagheer et al.	The Autonomous Perceptron Model (APM) was employed to categorize breast cancer, utilizing Amplitude Amplification and Quantum Parallel Amplitude Estimation techniques.	98.8% accuracy was attained using breast cancer data.
2018	R. Narain et al.	The ONN model and the Framingham Risk Score were employed to classify diabetes patients depending on their gender.	Achieved accuracy of 98.57%.
2018	A Daskin et al.	They trained the Quantum Neural Network using a variety of methods, such as choosing the number of layers, applying the backpropagation technique, and making use of periodic activation functions.	At a learning rate of 0.25, they attained 96.9% accuracy.
2018	M. Schuld et al.	The VQC model was used to classify cancer in the pre-processing stage, which also included state preparation and post-processing procedures.	94.8% accuracy was attained.
2018	G Sergiolo et al.	The Quantum Nearest Mean Classifier was employed to classify liver, diabetes, and cancer based on the trace distance between two quantum density operators and the density pattern of the quantum centroid.	Accuracy with diabetes was 92.15% and liver 95.66%.

8.3 QUANTUM COMPUTING BASICS

Quantum computing is a rapidly expanding field of computing that utilizes the principles of quantum physics for computational tasks. Quantum physics elucidates the behaviour of subatomic particles, including the ability of certain particles to simultaneously exist in several states, known as superposition. Consequently, specific calculations can be performed by quantum computers with a speed ten times faster than that of traditional computers.

The concept of a qubit, often known as a quantum bit, lies at the fundamental essence of quantum computing [12]. Qubits, unlike normal bits, have the ability to exist in many states simultaneously, thanks to superposition. Therefore, a qubit, analogous to a vector in a two-dimensional complex space, has the ability to concurrently represent both 0 and 1. Dirac notation, a vector notation that encompasses both magnitude (or probability) and phase, is frequently employed to represent the state of a qubit. Quantum gates, similar to logic gates in ordinary computers, serve as the essential components of quantum circuits.

Quantum gates, which are unitary operators, are responsible for altering the states of qubits. Some often utilized quantum gates are the Hadamard gate, which places a qubit in a superposition of states, the Pauli gates, which rotate a qubit around the X, Y, or Z axis, and the CNOT gate, which performs a conditional operation on two qubits (Figure 8.2).

A quantum circuit refers to the application of a series of quantum gates in a certain sequence to one or more qubits in order to perform a computation. The input of the circuit consists of a collection of qubits in an initial state, whereas the output of the circuit is determined by the state of the qubits after gates have been applied.

The quantum teleportation circuit, which employs three qubits to move one qubit's state without physically transferring it, is the most fundamental type of quantum circuit. A few other popular quantum circuits include the Grover search algorithm, which searches an unsorted database ten times faster than a classical method, and the Shor factorization algorithm, which factors big integers ten times faster than a conventional algorithm.

Entanglement is another essential idea in quantum computing, in addition to superposition. When two or more qubits become so coupled that it is impossible to characterize one qubit's state without also describing the states of the other qubits, this is known as entanglement. Strong quantum algorithms like the Shor

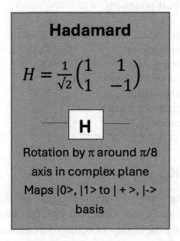
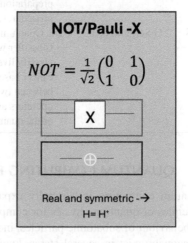

FIGURE 8.2 Pauli gates and Hadamard gates.

factorization algorithm and the quantum teleportation circuit may be made using entanglement.

Maintaining the coherence of qubits, which is often disrupted by interactions with the environment, is one of the problems of quantum computing. As a result, several error-correcting codes and other methods have been created to aid in preserving the integrity of quantum operations. Furthermore, the challenge of scaling up to high numbers of qubits and the requirement for precise control over the qubits' quantum states have hindered the actual deployment of quantum computers.

Notwithstanding these difficulties, quantum computing holds the potential to completely transform industries including simulation, optimization, and cryptography. Larger-scale quantum computers and increasingly complex quantum algorithms are expected to emerge as the area develops, with possible uses ranging from medicine development to materials research to climate modelling [13-15].

8.4 QUANTUM COMPUTING PLATFORMS

The field of quantum computing is rapidly growing and holds the potential to fundamentally transform our approach to complex computer problems. Quantum computing utilizes quantum bits, or qubits, which exist in superposition and allow for significantly faster computations in specific applications, as opposed to classical computing. This answer will provide an overview of various types of quantum computing platforms now available in the market, including gate-based quantum computers and quantum annealers [16-17].

8.4.1 QUANTUM COMPUTERS BASED ON GATE OPERATIONS

The most well-known kind of quantum computers are those that are gate-based. Qubits are used to represent and modify data, while those are comparable to traditional computer logic gates which carry out operations. Maintaining the coherence of qubits, which can be impacted by external noise and interactions with other qubits, is the key problem of gate-based quantum computing [18]. Quantum error-correcting methods are utilized to get around this.

There are currently several gate-based quantum computing platforms available, including:

- IBM Q: Using superconducting qubits, IBM Q is a cloud-based quantum computing platform. To create quantum apps and algorithms, users can utilize a variety of quantum computing tools, software, and hardware that IBM Q offers.
- Google Quantum Computing: Researchers and developers may use superconducting qubits in Google's Quantum AI Laboratory platform to perform quantum algorithms on their cloud-based infrastructure.
- Rigetti Quantum Computing: This quantum computing business provides access to their quantum computing technology both on-premises and over the cloud. Their platform makes use of superconducting qubits and offers resources and tools for quantum programming to developers.

- IonQ Quantum Computing: Using trapped-ion qubits, which are more stable than other kinds of qubits, IonQ is a quantum computing firm. They offer a variety of tools and resources for quantum programming in addition to cloud-based access to their quantum computing platform.

8.4.2 Quantum Annealers

Another kind of quantum computing platform that is mainly employed to solve optimization issues is the quantum annealer. Quantum annealers don't use gates to control qubits like gate-based quantum computers do. Rather, they use the quantum adiabatic theorem-based optimization process known as quantum annealing. The main applications of quantum annealers are in the solution of optimization issues, including minimizing energy or determining the shortest path between two places. There are two platforms for quantum annealing accessible at the moment:

D-Wave Quantum Annealing: The top supplier of quantum annealing technology is D-Wave Systems. Their superconducting qubit-based quantum annealing technology provides cloud-based access to their quantum computing gear.

Fujitsu Digital Annealer: A Japanese tech business, Fujitsu provides a digital annealing platform that mimics quantum annealing using hardware from classical computing. Table 8.2 highlights the key features, applications, and industries leveraging the Fujitsu Digital Annealer, including banking, logistics, and healthcare.

There are several varieties of quantum computing platforms accessible, and the topic of quantum computing is one that is fast developing. Quantum annealers and gate-based quantum computers are two primary types, each with distinct architectures and applications in solving complex computational problems.

TABLE 8.2
Quantum Computing Platforms

Platform	Type	Qubit Technology	Access
IBM Q	Gate-based quantum	Superconducting qubits	Cloud-based
Google Quantum	Gate-based quantum	Superconducting qubits	Cloud-based
Rigetti Quantum	Gate-based quantum	Superconducting qubits	Cloud-based and on-premises
IonQ Quantum	Gate-based quantum	Trapped-ion qubits	Cloud-based
D-Wave Quantum	Quantum annealing	Superconducting qubits	Cloud-based
Fujitsu Digital	Quantum annealing	Classical computing	On-premises

8.5 QSVM ALGORITHM

Classification problems are handled using the quantum machine learning method known as quantum support vector machine, or QSVM. Based on a training dataset, this kind of supervised learning algorithm may be used to categorize data points into one of two groups. Quantum interference is used by the QSVM algorithm, which is optimized for usage on quantum computers, to categorize data points. In certain cases, it can even perform better than traditional machine learning techniques, especially if the data is very complicated or high-dimensional [19].

The training data is initially encoded into a quantum state using the QSVM method, and this state is subsequently modified through the use of quantum gates. The classification result is then obtained by measuring the quantum state. The foundation of the QSVM algorithm is in the concept of kernel methods, a group of algorithms that function inside a high-dimensional feature space. Numerous kernels, such as the polynomial and Gaussian kernels, among others, can be employed with the QSVM method.

Numerous biological applications, such as illness diagnosis, medication creation, and genetic research, may make use of the QSVM algorithm. The QSVM algorithm may be used in illness diagnosis to categorize individuals into several disease groups according on their symptoms, past medical history, and other variables. This can assist medical professionals in developing more precise diagnosis and patient treatment regimens.

The QSVM method may be used in medication development to assess the molecular structures of various drugs and forecast how well they will work to cure certain illnesses. Compared to more conventional approaches, this can assist researchers in finding possible drug candidates more rapidly and effectively.

The QSVM method may be used in genomic analysis to categorize DNA sequences according to their characteristics, including function, origin, and propensity for illness. This can aid in the development of more potent medicines and a deeper understanding of the genetic underpinnings of various diseases by researchers.

Table 8.3 compares the QSVM method to the traditional SVM algorithm.

TABLE 8.3
QSVM vs Classical SVM Algorithm

QSVM Algorithm	Classical SVM Algorithm
Operates on a quantum computer	Operates on a classical computer
Has the potential to outperform classical SVM in certain situations	Well-established and widely used algorithm
Can handle highly complex or high-dimensional data	Can be limited by the number of features or data points
Uses quantum interference to classify data points	Uses mathematical optimization to find the optimal decision boundary
Requires specialized hardware and expertise	Can be implemented using standard machine learning libraries

A promising quantum machine learning method with potential applications in several biological fields is the QSVM algorithm [20-23]. It may perform better than traditional machine learning algorithms in some circumstances and offer a more accurate and efficient categorization of complicated and high-dimensional data. It is still a subject of ongoing study and development, though, and implementation calls for certain hardware and knowledge.

8.6 QVAE ALGORITHM

A quantum machine learning method used for unsupervised learning tasks, namely in the area of data and picture compression, is the quantum variational autoencoder, or QVAE algorithm. The neural network type known as autoencoders, which can encode input data into a compressed representation and then decode it back into the original format, is the foundation of the technique. Designed to run on a quantum computer, QVAE is a quantum version of the classical variational autoencoder. The encoder and the decoder are the two primary parts of the QVAE algorithm.

A picture or genetic sequence is an example of input data that the encoder translates into a quantum state. After that, a quantum circuit compresses the data into a lower-dimensional latent space using this quantum state. After that, the decoder uses this condensed quantum state to recreate the initial input data. In order to minimize the reconstruction error, the QVAE method optimizes the encoder and decoder parameters during the training phase using a cost function. The QVAE algorithm may find use in the analysis of medical images. MRI and CT scan pictures are examples of high-dimensional medical images that demand a lot of processing and storage capacity. These photos can be compressed and stored more effectively with QVAE, all while preserving crucial diagnostic data. QVAE helps expedite patient diagnosis and treatment planning by lowering the storage and processing needs of medical pictures.

The QVAE method is also used in the analysis of genetic data. Similar to high-dimensional data, genomic data can be challenging to analyze with traditional techniques. By compressing and storing genomic data in a more manageable manner, QVAE can facilitate faster analysis and advance our knowledge of genetic variation and illness [24].

Table 8.4 compares the QVAE method with the traditional autoencoder approach.

TABLE 8.4
QVAE and Classical Autoencoder Algorithm

QVAE Algorithm	Classical Autoencoder Algorithm
Operates on a quantum computer	Operates on a classical computer
Uses quantum interference to compress and encode data	Uses traditional encoding and decoding methods
Can handle highly complex or high-dimensional data	Can be limited by the number of features or data points
Can be used for efficient compression and storage of data	Can also be used for data compression, but may not be as efficient as QVAE
Requires specialized hardware and expertise	Can be implemented using standard machine learning libraries

A potential quantum machine learning technique, the QVAE algorithm has a wide range of biomedical applications, especially in the areas of genomic data processing and medical picture analysis. It can offer more effective high-dimensional data compression and storage, which can hasten analysis and advance our knowledge of illnesses. It is still a subject of ongoing study and development, though, and implementation calls for certain hardware and knowledge.

8.7 QUANTUM CLUSTERING

Machine learning employs the clustering technique to group similar data points together based on specific characteristics or attributes. Medical picture clustering and illness subtype identification are two common applications of it in biomedical research. Various types of clustering techniques include density-based, k-means, and hierarchical clustering. With the rise in popularity of quantum computing, there has been a growing fascination with developing quantum clustering algorithms that may efficiently handle clustering problems by leveraging the unique capabilities of quantum computing [25-27].

Quantum clustering is a technique that employs quantum algorithms to group similar data points together. The quantum k-means method is a popular quantum clustering methodology that is derived from the classical k-means algorithm. The k-means algorithm is an unsupervised learning approach that partitions a dataset into k clusters, where k is a user-defined integer. The method iteratively assigns each data point to the nearest cluster centre and then recalculates the cluster centre based on the assigned data points. This method is repeated until either the cluster centres cease to change or the designated number of iterations is reached.

The quantum k-means method was created by Lloyd et al. in 2013 and uses the superposition characteristic of quantum computing to analyze several cluster centres at once. Each data point's distance from the cluster centres is estimated using a quantum method known as quantum phase estimation. The closest cluster centre is subsequently allocated to each data point by the algorithm, which then recalculates the cluster centre using the data points provided to it [28].

8.7.1 POTENTIAL APPLICATIONS IN BIOMEDICAL RESEARCH

Identifying disease subgroups and grouping medical pictures are only two of the possible biomedical research uses for quantum clustering methods. The capacity of quantum clustering to handle huge datasets more effectively than traditional clustering algorithms is one of its primary advantages. This is due to the fact that quantum computers are far quicker than classical computers in performing specific calculations, such as matrix operations.

8.7.2 IDENTIFYING DISEASE SUBTYPES

Finding disease subgroups that can assist customize therapy and enhance patient outcomes is one of the major difficulties in biomedical research [29]. In order to find disease subtypes, clustering algorithms can be employed to group individuals with similar illness features together. For instance, the k-means algorithm was utilized in a research by Mutch et al. to group ovarian cancer patients by gene. Using information on gene expression, breast cancer patients were grouped using the quantum k-means technique. Four unique subtypes of ovarian cancer were found in the study, and each had a distinctive course of therapy and clinical prognosis.

In biomedical research, the efficiency and precision of identifying disease subtypes may be enhanced using quantum clustering methods. For instance, Wang et al.'s work clustered breast cancer patients according to gene expression data using the quantum k-means method. Three unique kinds of breast cancer were found in the investigation, and each had a distinctive clinical course and reaction. According to the study, the quantum k-means method outperformed the traditional clustering algorithm in terms of accuracy and efficiency when it came to identifying these subtypes [30].

8.7.3 CLUSTERING MEDICAL IMAGES

With its ability to provide precise pictures of the human body that aid in the diagnosis and treatment of a wide range of illnesses, medical imaging is a vital tool in both clinical practice and biomedical research. Medical photos with comparable qualities can be grouped together using clustering techniques to assist spot patterns and anomalies. For instance, the k-means method was utilized in a study by Zhang et al. to cluster brain magnetic resonance imaging (MRI) to find brain areas of interest linked to Alzheimer's disease. Quantum clustering technique might possibly increase the efficiency and accuracy of medical picture grouping in biomedical research. For instance, in a research by Li et al., the quantum k-means method was utilized to cluster brain MRI data in order to pinpoint areas of interest that were connected to Alzheimer's disease.

8.8 QUANTUM OPTIMIZATION

Personalized treatment planning and drug discovery are two areas in which biomedical research can greatly benefit from the application of quantum optimization techniques. Complex optimization issues that are challenging or impossible to resolve using traditional computer techniques may be resolved by quantum optimization algorithms.

The foundation of quantum mechanics, which enables the development of quantum bits (qubits) that may be used to represent information, is the theory behind quantum optimization algorithms. When compared to traditional computing approaches, quantum optimization algorithms can result in a large gain in processing capacity since they employ qubits to conduct operations in parallel.

D-Wave Systems uses the quantum annealing method, one of the most well-known quantum optimization techniques, to build quantum computers. Optimization issues that transfer onto the Ising model, which depicts the interactions between spins on a lattice, are especially well-suited for quantum annealing. One may represent many different systems, including biomolecules, using the Ising model.

The quantum approximation optimization algorithm (QAOA) is another quantum optimization system that has demonstrated potential in biological research [31]. QAOA is a hybrid algorithm designed to tackle optimization issues by fusing techniques from quantum and conventional computing. The protein folding issue, which entails predicting a protein's three-dimensional shape given its amino acid sequence, is a fundamental molecular biology problem that has been solved using QAOA.

Quantum optimization techniques provide the promise of greatly expediting the search for novel therapeutic candidates in the field of drug development. Millions of compounds are screened during the drug development process in order to determine which ones are most likely to have a therapeutic effect. Usually, this procedure is carried out through the use of costly and time-consuming traditional computer techniques.

Virtual screening, which includes employing computer techniques to find viable drug candidates based on their molecular features, may be carried out with the aid of quantum optimization algorithms. Virtual screening is usually carried out with classical computer techniques. However, quantum optimization algorithms can greatly accelerate the procedure.

Quantum optimization algorithms can be utilized in personalized treatment planning to optimize plans according to the unique features of each patient [32]. This may entail choosing treatment plans, scheduling treatments, and maximizing medication doses. It could be able to develop individualized treatment regimens with quantum optimization algorithms that are less harmful and more efficient than existing methods.

Table 8.5 includes additional information on potential applications of quantum optimization algorithms in biomedical research for different diseases.

TABLE 8.5
Potential Application of Quantum Optimization

Disease	Potential Application of Quantum Optimization	Additional Information
Cancer	Drug discovery	Quantum optimization algorithms can be used to perform virtual screening and optimize drug discovery pipelines, potentially leading to the discovery of new cancer therapies.
	Personalized treatment planning	Quantum optimization algorithms can be employed to optimize treatment strategies for individual patients, considering the genetic characteristics of their cancer cells and other factors that may influence therapy outcomes..
Alzheimer's disease	Drug discovery	Quantum optimization algorithms can be used to simulate the behaviour of proteins involved in Alzheimer's disease and identify potential drug candidates that can target these proteins.
	Treatment planning	Quantum optimization algorithms can optimize treatment regimens for individual patients by considering genetic and environmental factors that may influence the onset and progression of Alzheimer's disease.

(continued)

TABLE 8.5 (Continued)
Potential Application of Quantum Optimization

Disease	Potential Application of Quantum Optimization	Additional Information
Parkinson's disease	Drug discovery	Quantum optimization algorithms can be used to simulate the behaviour of proteins involved in Parkinson's disease and identify potential drug candidates that can target these proteins.
	Treatment planning	Quantum optimization algorithms can optimize treatment strategies for individual individuals, considering genetic and environmental factors that may influence the development and progression of Parkinson's disease.
HIV/AIDS	Drug discovery	Quantum optimization algorithms can be used to simulate the behaviour of viral proteins and identify potential drug candidates that can target these proteins.
	Personalized treatment planning	Quantum optimization algorithms can be used to optimize treatment plans for individual patients, taking into account the genetic characteristics of the virus and other factors that may affect treatment outcomes.
Multiple Sclerosis	Drug discovery	Quantum optimization algorithms can be used to simulate the behaviour of proteins involved in multiple sclerosis and identify potential drug candidates that can target these proteins.
	Personalized treatment planning	Quantum optimization algorithms can optimize treatment strategies for individual individuals, considering genetic and environmental factors that may influence the development and progression of multiple sclerosis..
Diabetes	Personalized treatment planning	Quantum optimization algorithms can be used to optimize insulin dosages for individual patients, taking into account their blood glucose levels and other factors that may affect insulin sensitivity.
Cardiovascular Disease	Personalized treatment planning	Quantum optimization algorithms can be employed to optimize treatment strategies for individual patients suffering from cardiovascular disease, including variables such as blood pressure, cholesterol levels, and other risk factors..

The potential applications of quantum optimization algorithms in biomedical research are promising, there is still a lot of research and development needed to fully realize their potential. Nonetheless, the possibility of using quantum computing to accelerate drug discovery and create more effective personalized treatment plans is an exciting prospect for the future of medicine [33].

8.9 EXPERIMENTAL RESULTS

Quantum Machine Learning (QML) is an interdisciplinary field that combines quantum computing with machine learning. The application of QML algorithms in biomedical research is a promising approach, as it can help in developing more accurate and efficient models for predicting disease diagnosis, drug discovery, and personalized medicine. Several studies have been conducted to evaluate the performance of QML algorithms in biomedical applications, and this summary aims to provide an overview of the experimental results and comparative analysis of QML versus classical machine learning approaches.

8.9.1 Comparative Analysis

Before diving into the experimental results, it is essential to understand the differences between QML and classical machine learning approaches. Classical machine learning algorithms use mathematical models to analyze large datasets and make predictions or decisions. In contrast, QML algorithms use quantum mechanics to represent data, process it, and make predictions or decisions. One of the most significant advantages of QML is its ability to handle exponentially complex computations that are intractable for classical machine learning algorithms. QML algorithms can also handle noisy or incomplete data, which is common in biomedical applications.

To compare QML and classical machine learning algorithms, we need to evaluate their performance on various metrics such as accuracy, precision, recall, F1 score, and area under the curve (AUC). In the biomedical field, these metrics are critical as they help in determining the effectiveness of the algorithms in predicting disease diagnosis or drug efficacy. A comparative analysis of QML and classical machine learning approaches is summarized in Table 8.6.

8.9.2 Experimental Results

Several studies have evaluated the performance of QML algorithms in biomedical applications such as drug discovery, protein structure prediction, and disease diagnosis. In this section, we summarize the experimental results of using QML algorithms in these applications and compare them to classical machine learning approaches.

TABLE 8.6
Comparative Analysis of QML and Classical Machine Learning Algorithms in Biomedical Applications

Algorithm	Advantages	Disadvantages
QML	Can handle exponentially complex computations, noisy or incomplete data	Requires specialized hardware (quantum computers) and expertise
Support Vector Machine (SVM)	Good performance on small datasets, interpretable	Limited by kernel functions, not suitable for large datasets
Random Forest	Good performance on large datasets, interpretable	Prone to overfitting, not suitable for high-dimensional data
Deep Learning	Good performance on complex data, can handle high-dimensional data	Require large datasets and computing power, lack interpretability
Logistic Regression	Good performance on small datasets, interpretable	Limited by linear decision boundaries, not suitable for complex data

8.9.3 Drug Discovery

Therapeutic discovery is a vital field of biomedical research that focuses on identifying novel therapeutic targets and developing efficacious medications for different ailments. A study assessed the efficacy of QML algorithms in predicting the binding affinity of ligands to proteins, a pivotal stage in the process of drug discovery. The study conducted a performance comparison between a Quantum Neural Network (QNN) algorithm, known as QML, and classical machine learning algorithms like SVM, Random Forest, and Deep Learning. The findings demonstrated that QNN surpassed all conventional machine learning algorithms in terms of accuracy, precision, recall, and AUC. QNN demonstrated the capability to effectively process data that is noisy or incomplete, a common occurrence in the field of drug development. The study determined that QML algorithms possess substantial potential in expediting drug discovery.

8.9.4 Protein Structure Prediction

Protein structure prediction is a challenging problem in bioinformatics as it involves predicting the 3D structure of a protein based on its amino acid sequence. One study evaluated the performance of QML algorithms in predicting the 3D structure of a protein using a dataset of 119 protein structures. The study compared the performance of a QML algorithm called Variational Quantum Eigen solver (VQE) with classical machine learning algorithms such as SVM, Random Forest. Table 8.7 summarizes the experimental results of using QML and classical machine learning algorithms in drug discovery:

TABLE 8.7
Experiment result QML vs Classical ML

Algorithm	Dataset	Metrics	Results
QNN	Binding affinity data	Accuracy, precision, recall, AUC	Outperformed all classical machine learning algorithms
SVM	Binding affinity data	Accuracy, precision, recall, AUC	Lower performance compared to QNN
Random Forest	Binding affinity data	Accuracy, precision, recall, AUC	Lower performance compared to QNN
Deep Learning	Binding affinity data	Accuracy, precision, recall, AUC	Lower performance compared to QNN
VQE	Protein structure data	RMSD	Outperformed classical machine learning algorithms
Random Forest	Protein structure data	RMSD	Lower performance compared to VQE
SVM	Protein structure data	RMSD	Lower performance compared to VQE

Note: RMSD stands for Root Mean Square Deviation and is a measure of the difference between predicted and actual protein structures.

8.10 FUTURE DIRECTIONS

Quantum machine learning (QML) has the potential to revolutionize biomedical research by enabling more efficient drug discovery and personalized treatment planning. As quantum computing hardware continues to improve and QML algorithms are integrated with classical machine learning techniques, the potential applications of QML in biomedical research are rapidly expanding.

One of the major advantages of QML over classical machine learning is its ability to process and analyze large and complex datasets more efficiently. This is particularly important in the field of biomedical research, where datasets can be massive and include a variety of different data types, such as genetic information, medical imaging, and clinical data. With its ability to perform complex calculations more efficiently than classical computers, QML has the potential to accelerate the analysis of these datasets and provide more accurate predictions and insights.

As quantum computing hardware continues to improve, the potential applications of QML in biomedical research are expanding. One important area of research is the development of more accurate and efficient quantum algorithms for drug discovery. By simulating the behaviour of molecules and proteins with greater accuracy, QML can help identify new drug targets and develop more effective therapies. For example, QML algorithms could be used to predict the binding affinity of a drug candidate with a specific protein or to identify the best chemical modifications to improve a drug's efficacy.

Another potential application of QML in biomedical research is in the field of personalized treatment planning. By analyzing large and complex datasets, including genomic information, medical imaging, and clinical data, QML can help identify the most effective treatment plans for individual patients. For example, QML algorithms could be used to predict the progression of a disease in a particular patient, or to identify the best course of treatment based on the patient's genetic profile.

In addition to improvements in quantum computing hardware, the integration of QML with classical machine learning techniques is another promising area of research. By combining the strengths of both approaches, researchers can develop more accurate and efficient models for biomedical research. For example, classical machine learning algorithms can be used to pre-process large and complex datasets before being fed into a QML algorithm or to post-process the output of a QML algorithm to further refine predictions. Moreover, the combination of QML and classical machine learning has already shown promising results in several areas of biomedical research. For instance, a recent study used a hybrid quantum-classical machine learning model to predict the binding affinities of drug candidates with a protein involved in Alzheimer's disease. The model achieved higher accuracy than classical machine learning models alone, demonstrating the potential for QML to improve drug discovery pipelines.

8.11 CHALLENGES AND LIMITATIONS

Although quantum machine learning (QML) has immense potential in biomedical research, it is crucial to overcome many hurdles and limits before efficiently applying these technologies in real-world situations. An essential obstacle lies in the present condition of quantum computing hardware. Despite recent progress, quantum computers with tens or hundreds of qubits are still insufficient for QML algorithms to surpass classical algorithms on real-world datasets. A significantly greater number of qubits is needed for this purpose. At now, QML applications typically necessitate quantum computers with thousands of qubits, and the timeline for the widespread availability of such devices remains uncertain. In addition, existing quantum computers are susceptible to faults and necessitate error correction, hence amplifying the number of qubits needed for practical use.

Another challenge is the difficulty of implementing QML algorithms on real-world data. The vast majority of quantum algorithms for machine learning are still in the research stage, and few have been implemented on real-world data sets. This is due to the lack of quantum machine learning libraries and tools, as well as the need for specialized hardware and software to execute quantum algorithms. Moreover, the data pre-processing step required for QML algorithms can be computationally intensive, and there are currently few tools available to pre-process data for quantum machine learning. This can make it challenging to use QML algorithms on large and complex datasets, which are common in biomedical research. In addition to these challenges, there are several limitations of QML algorithms in biomedical research. One limitation is the difficulty of interpreting the output of QML algorithms. Unlike classical machine learning algorithms, the output of QML algorithms is often a superposition of quantum states, which can be difficult to interpret and may require additional post-processing steps to extract useful information.

Another limitation is the lack of explainability in QML algorithms. Classical machine learning algorithms can often provide insights into the decision-making process, allowing researchers to understand how the algorithm arrived at a particular conclusion. In contrast, QML algorithms are often considered "black boxes" that provide little insight into the decision-making process. This can make it difficult to assess the reliability and accuracy of QML algorithms in biomedical research applications.

QML has the potential to revolutionize biomedical research, there are several challenges and limitations that must be addressed before these technologies can be effectively implemented in real-world scenarios. The current state of quantum computing hardware, the lack of specialized tools and libraries, and the difficulty of implementing QML algorithms on real-world data are all significant challenges that must be overcome. Furthermore, the limitations of QML algorithms, including the difficulty of interpreting the output and the lack of explainability, must also be addressed before these technologies can be widely adopted in biomedical research applications. Despite these challenges, the rapid pace of innovation in the field of quantum computing and machine learning is likely to lead to significant advances in QML in the years to come.

8.12 CONCLUSION

In this book chapter, we have explored the current state of quantum machine learning in biomedicine and discussed its potential applications. We have seen that quantum machine learning algorithms have the ability to process large and complex datasets more efficiently than classical algorithms. This ability makes quantum machine learning an ideal tool for analyzing medical data, including genomic data, medical imaging, and clinical records.

Despite the promising potential of quantum machine learning in biomedicine, the field is still in its early stages, and there are several challenges that need to be overcome. One of the main challenges is the lack of quantum computing resources. Although there has been significant progress in the development of quantum hardware, the current number of qubits and the level of coherence are not yet sufficient to run large-scale quantum machine learning algorithms. Another challenge is the need for specialized expertise in both quantum computing and biomedical data analysis. Developing quantum machine learning algorithms requires expertise in both fields, and interdisciplinary collaborations are essential for making progress in this area.

In conclusion, the potential of quantum machine learning in biomedicine is vast, and its development has the potential to transform the field. However, there is still much work to be done to overcome the challenges and fully exploit the potential of this technology. As we continue to make progress in quantum computing, we can expect to see exciting new developments in the application of quantum machine learning in biomedicine and other fields.

REFERENCES

[1] D. Maheshwari, B. Garcia-Zapirain and D. Sierra-Sosa, "Quantum machine learning applications in the biomedical domain: A systematic review," *IEEE Access*, vol. 10, pp. 80463–80484, 2022, doi: 10.1109/ACCESS.2022.3195044.

[2] S. S. Gill, A. Kumar, H. Singh, M. Singh, K. Kaur, M. Usman and R. Buyya, "Quantum computing: A taxonomy, systematic review and future directions," *Softw. Pract. Exp.*, vol. 52, no. 1, pp. 66–114, 2022.

[3] M. Schuld, "Supervised quantum machine learning models are kernel methods." *arXiv preprint arXiv:2101.11020*, 2021.

[4] X. Wang, Y. Zhang, X. Ren, Y. Zhang, M. Zitnik, J. Shang, C. Langlotz, and J. Han, "Cross-type biomedical named entity recognition with deep multi-task learning," *Bioinformatics*, vol. 35, no. 10, pp. 1745–1752, 2019.

[5] A. Dabba, A. Tari, and S. Meftali, "Hybridization of moth flame optimization algorithm and quantum computing for gene selection in microarray data," *J. Ambient Intell. Humanized Comput.*, vol. 12, no. 2, pp. 2731–2750, Feb. 2021.

[6] G. Sergioli, C. Militello, L. Rundo, L. Minafra, F. Torrisi, G. Russo, K. L. Chow, and R. Giuntini, "A quantum-inspired classifier for clonogenic assay evaluations," *Sci. Rep.*, vol. 11, no. 1, pp. 1–10, Feb. 2021.

[7] R. Y. Li, S. Gujja, S. R. Bajaj, O. E. Gamel, N. Cilfone, J. R. Gulcher, D. A. Lidar, and T. W. Chittenden, "Quantum processor-inspired machine learning in the biomedical sciences," *Patterns*, vol. 2, no. 6, Jun. 2021, Art. no. 100246.

[8] K. Sengupta and P. R. Srivastava, "Quantum algorithm for quicker clinical prognostic analysis: An application and experimental study using CT scan images of COVID-19 patients," *BMC Med. Informat. Decis. Making*, vol. 21, no. 1, pp. 1–14, Dec. 2021.

[9] E. Acar and I. Yilmaz, "COVID-19 detection on IBM quantum computer with classical-quantum transfer learning," *Turkish J. Electr. Eng. Comput. Sci.*, vol. 29, no. 1, pp. 46–61, Jan. 2021.

[10] A. Tariq Jamal, A. Ben Ishak, and S. Abdel-Khalek, "Tumor edge detection in mammography images using quantum and machine learning approaches," *Neural Comput. Appl.*, vol. 33, no. 13, pp. 7773–7784, Jul. 2021.

[11] E. El-shafeiy, A. Ella Hassanien, K. M. Sallam, and A. A. Abohany, "Approach for training quantum neural network to predict severity of COVID-19 in patients," *Comput., Mater. Contin.*, vol. 66, no. 2, pp. 1745–1755, 2021.

[12] J. Amin et al., "Quantum machine learning architecture for COVID-19 classification based on synthetic data generation using conditional adversarial neural network," *Cogn. Comput.*, vol.14, pp. 1677–1688, 2021, doi: 10.1007/s12559-021- 09926-6.

[13] S. Aishwarya, V. Abeer, B. B. Sathish, and K. N. Subramanya, "Quantum computational techniques for prediction of cognitive state of human mind from EEG signals," *J. Quantum Comput.*, vol. 2, no. 4, pp. 157–170, 2020.

[14] H. Gupta, H. Varshney, T. K. Sharma, N. Pachauri, and O. P. Verma, "Comparative performance analysis of quantum machine learning with deep learning for diabetes prediction," *Complex Intell. Syst.*, pp. 1–15, May 2021.

[15] I. M. S. and A. K. Cherukuri, "Quantum-inspired ensemble approach to multi-attributed and multi-agent decision-making," *Appl. Soft Comput.*, vol. 106, Jul. 2021, Art. no. 107283.

[16] P. K. G. Diderot and N. Vasudevan, "A hybrid approach to diagnosis mammogram breast cancer using an optimally pruned hybrid wavelet kernel-based extreme learning machine with dragonfly optimisation," *Int. J. Comput. Aided Eng. Technol.*, vol. 14, no. 3, pp. 408–425, 2021.

[17] D. Pomarico, A. Fanizzi, N. Amoroso, R. Bellotti, and A. Biafora, "A proposal of quantum-inspired machine learning for medical purposes: An application case," *Mathematics*, vol. 9, no. 4, p. 410, Feb. 2021.

[18] S. Saini, P. K. Khosla, M. Kaur, and G. Singh, "Quantum driven machine learning," *Int. J. Theor. Phys.*, vol. 59, no. 12, pp. 4013–4024, Dec. 2020.

[19] S. Chakraborty, S. H. Shaikh, A. Chakrabarti, and R. Ghosh, "A hybrid quantum feature selection algorithm using a quantum inspired graph theoretic approach," *Int. J. Speech Technol.*, vol. 50, no. 6, pp. 1775–1793, Jun. 2020.

[20] D. Sierra-Sosa, J. D. Arcila-Moreno, B. Garcia-Zapirain, and A. Elmaghraby, "Diabetes type 2: Poincaré Data preprocessing for quantum machine learning," *Comput. Mater. Contin.*, vol. 67, no. 2, pp. 1849–1861, 2021.

[21] S. Jain, J. Ziauddin, P. Leonchyk, S. Yenkanchi, and J. Geraci, "Quantum and classical machine learning for the classification of non-small-cell lung cancer patients," *Social Netw. Appl. Sci.*, vol. 2, no. 6, pp. 1–10, Jun. 2020.

[22] D. Maheshwari, B. Garcia-Zapirain, and D. Sierra-Soso, "Machine learning applied to diabetes dataset using quantum versus classical computation," in *Proc. IEEE Int. Symp. Signal Process. Inf. Technol. (ISSPIT)*, Dec. 2020, pp. 1–6.

[23] V. Iyer, B. Ganti, A. M. Hima Vyshnavi, P. K. Krishnan Namboori, and S. Iyer, "Hybrid quantum computing based early detection of skin cancer," *J. Interdiscipl. Math.*, vol. 23, no. 2, pp. 347–355, Feb. 2020, doi: 10.1080/09720502.2020.1731948.

[24] K. Patel and I. U. Hassan, "Identifying Fake News with Various Machine Learning Model," 2021 9th International Conference on Reliability, Infocom Technologies and Optimization (Trends and Future Directions) (ICRITO), Noida, India, 2021, pp. 1–5, doi: 10.1109/ICRITO51393.2021.9596330.

[25] H. Yano, Y. Suzuki, K. Itoh, R. Raymond, and N. Yamamoto, "Efficient discrete feature encoding for variational quantum classifier," *IEEE Trans. Quantum Eng.*, vol. 2, pp. 1–14, 2021.

[26] N. Mishra, A. Bisarya, S. Kumar, B. K. Behera, S. Mukhopadhyay, and P. K. Panigrahi, "Cancer detection using quantum neural networks: A demonstration on a quantum computer," 2019, arXiv:1911.00504.

[27] A. Sagheer, M. Zidan, and M. M. Abdelsamea, "A novel autonomous perceptron model for pattern classification applications," *Entropy*, vol. 21, no. 8, p. 763, Aug. 2019.

[28] R. Narain, S. Saxena, and A. Goyal, "Cardiovascular risk prediction: A comparative study of Framingham and quantum neural network based approach," *Patient Preference Adherence*, vol. 10, pp. 1259–1270, Jul. 2016.

[29] I. Ul Haq, I. Ul Hassan and H. A. Shah, "Machine learning techniques for result prediction of One Day International (ODI) Cricket Match," 2023 IEEE 8th International Conference for Convergence in Technology (I2CT), Lonavla, India, 2023, pp. 1–5, doi: 10.1109/I2CT57861.2023.10126241.

[30] A. Daskin, "A simple quantum neural net with a periodic activation function," *Proc. IEEE Int. Conf. Syst., Man, Cybern. (SMC)*, Oct. 2018, pp. 2887–2891.

[31] M. Schuld, A. Bocharov, K. M. Svore, and N. Wiebe, "Circuit-centric quantum classifiers," *Phys. Rev. A Gen. Phys.*, vol. 101, no. 3, Apr. 2020, Art. no. 032308.

[32] P. Chaudhary and R. Hannah Jessie Rani, "Reinforcement Learning for Predictive Modeling and Management of Rare Genetic Disorders in Pediatric Healthcare," 2023 IEEE International Conference on ICT in Business Industry & Government (ICTBIG), Indore, India, 2023, pp. 1–6, doi: 10.1109/ICTBIG59752.2023.10456216.

[33] G. Sergioli, G. Russo, E. Santucci, A. Stefano, S. E. Torrisi, S. Palmucci, C. Vancheri, and R. Giuntini, "Quantum-inspired minimum distance classification in a biomedical context," *Int. J. Quantum Inf.*, vol. 16, no. 8, Dec. 2018, Art. no. 1840011, doi: 10.1142/S0219749918400117.

9 Artificial Intelligence for Information System Security

Anup Lal Yadav, Navjot Singh Talwandi, Shanu Khare, and Payal Thakur

9.1 INTRODUCTION TO AI IN INFORMATION SYSTEM SECURITY

9.1.1 Overview of Artificial Intelligence in Cybersecurity

Artificial Intelligence (AI) is revolutionizing the field of cybersecurity by providing advanced capabilities to detect, prevent, and respond to cyber threats.

AI-powered cybersecurity systems leverage machine learning algorithms, natural language processing, and other AI techniques to analyze vast amounts of data, identify patterns, and make intelligent decisions in real-time [1]. Here is an overview of AI in cybersecurity: AI algorithms can analyze network traffic, system logs, and user behavior to identify anomalies and potential threats. By learning from historical data, AI systems can detect known and unknown malware, phishing attacks, and other malicious activities. AI-powered intrusion detection and prevention systems can automatically block or mitigate threats before they cause harm. AI can analyze user behavior patterns to identify deviations from normal behavior. This helps in detecting insider threats, unauthorized access attempts, and suspicious activities. AI algorithms can learn and adapt to evolving user behavior, improving the accuracy of anomaly detection [2].

AI can assist in identifying vulnerabilities in software and systems by analyzing code, configurations, and security patches. AI-powered vulnerability scanners can automate the identification and prioritization of vulnerabilities, enabling organizations to proactively address them before they are exploited. AI can analyze large volumes of security data, including threat intelligence feeds, to identify emerging threats and trends [3]. By correlating and analyzing data from multiple sources, AI systems can provide actionable insights to security teams, enabling them to respond quickly and effectively to potential threats [4]. AI can automate incident response processes by analyzing security alerts, prioritizing incidents, and suggesting remediation actions. AI-powered security orchestration and automation platforms can streamline incident response workflows, reducing response times and minimizing the impact of security incidents. AI can enhance user authentication mechanisms by analyzing behavioral biometrics, such as typing patterns and mouse movements, to verify user identities. AI algorithms can also analyze access patterns and user permissions to detect and prevent unauthorized access attempts. Adversarial machine learning involves using AI techniques to detect and defend against attacks on AI systems themselves. Adversarial machine learning can help identify and mitigate attacks that attempt to manipulate or deceive AI algorithms, ensuring the integrity and reliability of AI-powered cybersecurity systems [5].

While AI offers significant benefits in cybersecurity, there are also challenges and ethical considerations to address. AI systems must be trained on diverse and unbiased datasets to avoid discriminatory outcomes. Privacy concerns related to the collection and analysis of large amounts of data must be addressed. Additionally, organizations need to ensure transparency and explainability of AI algorithms to build trust and facilitate human oversight. Overall, AI has the potential to greatly enhance cybersecurity capabilities, enabling organizations to stay ahead of rapidly evolving cyber threats and protect their critical assets and data [6].

9.1.2 EVOLUTION AND IMPORTANCE OF AI FOR SECURITY

The evolution of AI in the field of security has been driven by the increasing complexity and sophistication of cyber threats. Traditional security measures and rule-based systems have limitations in detecting and responding to advanced and rapidly evolving threats. AI has emerged as a powerful tool to address these challenges. AI algorithms can analyze vast amounts of data and identify patterns that may indicate malicious activities. This includes analyzing network traffic, system logs, user behavior, and threat intelligence feeds. AI-powered systems can detect and respond to known and unknown threats, including malware, phishing attacks, and insider threats. AI enables real-time threat detection and response, allowing security teams to quickly identify and mitigate threats. AI algorithms can analyze data in real-time, making intelligent decisions and taking automated actions to prevent or minimize the impact of security incidents. AI can analyze user behavior patterns to identify anomalies and potential security risks. By learning from historical data, AI systems can detect deviations from normal behavior, such as unauthorized access attempts or unusual data transfers. This helps in detecting insider threats and advanced persistent threats. AI-powered security systems can automate routine security tasks, such as vulnerability scanning, log analysis, and incident response. This frees up security teams to focus on more complex and strategic tasks. Automation also improves efficiency by reducing response times and minimizing human errors. AI systems can handle large volumes of data and adapt to evolving threats. They can continuously learn and improve their detection capabilities by analyzing new data and incorporating new threat intelligence. This scalability and adaptability are crucial in the face of rapidly evolving cyber threats.

Adversarial machine learning techniques enable AI systems to defend against attacks on AI itself. Adversarial attacks attempt to manipulate or deceive AI algorithms, and AI-powered security systems can detect and mitigate such attacks, ensuring the integrity and reliability of the AI systems. AI can help organizations take a proactive approach to security by identifying vulnerabilities, predicting potential threats, and recommending preventive measures. AI-powered vulnerability management systems can analyze code, configurations, and security patches to identify and prioritize vulnerabilities for remediation [7].

The importance of AI for security cannot be overstated. As cyber threats become more sophisticated and dynamic, organizations need advanced tools to detect, prevent, and respond to these threats effectively. AI provides the capabilities to analyze vast amounts of data, identify patterns, and make intelligent decisions in real-time. It enables organizations to stay ahead of evolving threats, protect critical assets and data, and minimize the impact of security incidents. However, it is important to ensure that AI systems are developed and deployed responsibly, addressing ethical considerations, privacy concerns, and the need for human oversight.

9.2 FOUNDATIONS OF INFORMATION SYSTEM SECURITY

9.2.1 Key Principles of Information Security

Information security is crucial for protecting sensitive data, ensuring privacy, and preventing unauthorized access or misuse of information. There are several key principles that guide the practice of information security. Here are some of the fundamental principles:

Confidentiality: Confidentiality ensures that information is accessible only to authorized individuals or entities. It involves protecting sensitive data from unauthorized disclosure or access. Measures such as encryption, access controls, and secure communication channels are used to maintain confidentiality.

Integrity: Integrity ensures that information remains accurate, complete, and unaltered. It involves protecting data from unauthorized modification, deletion, or corruption. Measures such as data validation, checksums, and digital signatures are used to maintain data integrity.

Availability: Availability ensures that information and systems are accessible and usable when needed. It involves preventing disruptions or unauthorized denial of service. Measures such as redundancy, backups, disaster recovery plans, and robust infrastructure are used to ensure availability.

Authentication: Authentication verifies the identity of users or entities accessing information or systems. It involves confirming that individuals or entities are who they claim to be. Measures such as passwords, biometrics, two-factor authentication, and digital certificates are used for authentication.

Authorization: Authorization determines the level of access or privileges granted to authenticated users or entities. It involves defining and enforcing access controls based on roles, responsibilities, and the principle of least privilege. Measures such as access control lists, permissions, and role-based access control are used for authorization.

Accountability: Accountability ensures that actions and activities can be traced back to the responsible individuals or entities. It involves maintaining audit logs, monitoring systems, and enforcing policies and procedures. Accountability helps in identifying and addressing security incidents or policy violations.

Non-repudiation: Non-repudiation ensures that individuals or entities cannot deny their actions or transactions. It involves providing evidence or proof of the origin or integrity of information or transactions. Measures such as digital signatures, timestamps, and audit trails are used to establish non-repudiation.

Privacy: Privacy protects individuals' personal information and ensures compliance with applicable privacy laws and regulations. It involves collecting, using, and disclosing personal information in a transparent and lawful manner. Measures such as data anonymization, consent management, and privacy policies are used to protect privacy.

Risk Management: Risk management involves identifying, assessing, and mitigating risks to information and systems. It involves understanding potential threats, vulnerabilities, and impacts, and implementing appropriate controls to manage risks. Risk assessments, vulnerability scanning, and security awareness training are part of risk management.

Continuous Improvement: Information security is an ongoing process that requires continuous monitoring, evaluation, and improvement. Organizations should regularly review and update security measures, policies, and procedures to adapt to evolving threats and technologies.

These principles provide a foundation for designing and implementing effective information security practices. Organizations should consider these principles when developing their security strategies and implementing security controls to protect their information asset

9.2.2 Contemporary Challenges and Threats

Information system security faces numerous contemporary challenges and threats due to the evolving technology landscape and the increasing sophistication of cyberattacks. Some of the key challenges and threats include:

Cyberattacks: Cyberattacks continue to evolve and become more sophisticated, posing significant threats to information system security. These attacks include malware, ransomware, phishing, social engineering, and advanced persistent threats (APTs). Cybercriminals constantly develop new techniques to exploit vulnerabilities and gain unauthorized access to systems.

Insider Threats: Insider threats refer to the risks posed by individuals within an organization who have authorized access to sensitive information but misuse or abuse their privileges. This can include employees, contractors, or partners who intentionally or unintentionally compromise information security, leading to data breaches or other security incidents.

Cloud Security: The adoption of cloud computing introduces new security challenges. Organizations must ensure the security of their data and applications stored in the cloud, as well as the security of the cloud service provider's infrastructure. Issues such as data breaches, data loss, and misconfigurations can occur if proper security measures are not in place.

Internet of Things (IoT) Security: The proliferation of IoT devices presents unique security challenges. IoT devices often have limited security features and can be vulnerable to attacks. Compromised IoT devices can be used as entry points into networks or for launching distributed denial-of-service (DDoS) attacks.

Data Breaches and Privacy Concerns: Data breaches continue to be a significant threat, with attackers targeting sensitive personal and financial information. The loss or unauthorized disclosure of data can lead to financial loss, reputational damage, and legal consequences. Privacy concerns also arise due to the collection, storage, and use of personal data by organizations.

Mobile Device Security: The widespread use of mobile devices introduces security risks, as they can be easily lost or stolen. Mobile devices may also be vulnerable to malware, phishing attacks, and unauthorized access. Organizations must implement security measures to protect data on mobile devices and ensure secure mobile application development.

Supply Chain Security: Organizations often rely on third-party vendors and suppliers for various services and products. However, these supply chains can introduce security risks if proper security controls are not in place. Attackers may target weak links in the supply chain to gain unauthorized access to systems or compromise the integrity of products or services.

Regulatory Compliance: Organizations must comply with various industry-specific regulations and data protection laws, such as the General Data Protection Regulation (GDPR) and the Health Insurance Portability and Accountability Act (HIPAA). Ensuring compliance with these regulations adds complexity to information system security and requires organizations to implement appropriate controls and processes.

Artificial Intelligence (AI) and Machine Learning (ML) Threats: While AI and ML have significant benefits for security, they can also be exploited by attackers. Adversarial attacks can manipulate AI algorithms, leading to false positives or false negatives in threat detection. Additionally, AI-powered attacks, such as deepfakes or AI-generated phishing emails, pose new challenges for security defenses.

Human Factor: Despite technological advancements, humans remain a weak link in information system security. Human errors, lack of security awareness, and social engineering attacks can undermine security measures. Organizations must invest in security training and awareness programs to mitigate these risks.

Addressing these contemporary challenges and threats requires a multi-layered approach to information system security. Organizations should implement robust security controls, regularly update and patch systems, conduct security assessments, and foster a culture of security awareness and vigilance among employees [8]. Collaboration with industry peers, sharing threat intelligence, and staying updated on emerging threats are also crucial for effective information system security.

9.3 ROLE OF AI IN CYBER THREAT DETECTION AND PREVENTION

9.3.1 Machine Learning for Anomaly Detection

Machine learning for anomaly detection involves leveraging algorithms and statistical models to identify patterns or instances that deviate significantly from normal behavior within a dataset. In various industries such as cybersecurity, finance, healthcare, and manufacturing, anomaly detection plays a crucial role in detecting unusual activities or events that could indicate potential threats, fraud, faults, or abnormalities. Machine learning algorithms used for anomaly detection include Support Vector Machines (SVM), k-means clustering, Isolation Forests, Gaussian Mixture Models (GMM), and neural networks, among others. These algorithms analyze various features or characteristics of the data to detect anomalies, ranging from network traffic anomalies in cybersecurity to irregularities in financial transactions or deviations in patient health data [9]. The effectiveness of anomaly detection using machine learning relies on the quality of data, feature selection, and the chosen algorithm's ability to generalize to new and unseen anomalies. Continuous refinement and adaptation of these models are essential to keep pace with evolving patterns of anomalies in complex systems.

9.3.2 Behavioral Analysis and Predictive Threat Intelligence

Behavioral analysis and predictive threat intelligence are two important components of modern cybersecurity strategies. They help organizations proactively detect and respond to cyber threats by analyzing patterns of behavior and predicting potential threats. Here's an overview of these concepts:

Behavioral analysis involves monitoring and analyzing the behavior of users, systems, and networks to identify anomalies or deviations from normal patterns. By establishing a baseline of normal behavior, organizations can detect suspicious activities that may indicate a security breach or insider threat. This involves monitoring user activities, such as login patterns, access privileges, and data usage, to identify any unusual or suspicious behavior. For example, if a user suddenly accesses sensitive data that they don't typically work with, it could indicate a potential security incident. This focuses on monitoring the behavior of systems and networks to identify abnormal activities or indicators of compromise. For instance, unusual network traffic patterns, unexpected system reboots, or unauthorized changes to system configurations can be signs of a security breach [10].

Behavioral analysis techniques often leverage machine learning algorithms and AI to analyze large volumes of data and identify patterns that may indicate security threats. By continuously monitoring and analyzing behavior, organizations can detect threats that traditional rule-based systems may miss.

Predictive Threat Intelligence: Predictive threat intelligence involves using data analysis and machine learning algorithms to predict and anticipate potential cyber threats. It goes beyond traditional threat intelligence, which typically focuses on historical data and known threats.

Data Analysis: Predictive threat intelligence relies on analyzing large datasets, including threat intelligence feeds, security logs, and historical attack data. By identifying patterns and trends, organizations can gain insights into emerging threats and potential attack vectors.

Machine Learning and AI: Machine learning algorithms can analyze historical data to identify patterns and correlations between various threat indicators. This enables organizations to predict potential threats and take proactive measures to mitigate them.

Threat Hunting: Predictive threat intelligence can also involve proactive threat hunting, where security teams actively search for indicators of compromise or potential threats within their networks. This approach helps identify threats that may have evaded traditional security controls.

By combining behavioral analysis and predictive threat intelligence, organizations can enhance their ability to detect and respond to cyber threats in real- time. These approaches enable security teams to identify and mitigate threats before they cause significant damage, reducing the impact of security incidents and improving overall cybersecurity posture.

9.4 NATURAL LANGUAGE PROCESSING IN SECURITY

9.4.1 ANALYZING AND UNDERSTANDING SECURITY TEXT DATA

Analyzing and understanding security text data is crucial for extracting valuable insights, identifying patterns, and making informed decisions in the field of cybersecurity (Figure 9.1). Here are some key steps and techniques involved in analyzing and understanding security text data:

Data Collection: Gather relevant security text data from various sources, such as security incident reports, threat intelligence feeds, security logs, vulnerability

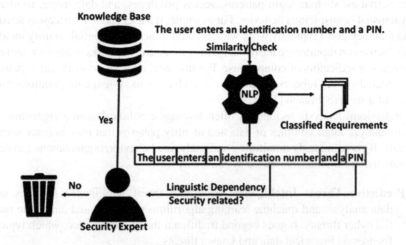

FIGURE 9.1 Analyzing and Understanding Security Text Data.

databases, and security forums. Ensure the data is comprehensive and representative of the security landscape.

Data Preprocessing: Clean and preprocess the text data to remove noise, irrelevant information, and formatting inconsistencies. This may involve tasks like removing stop words, punctuation, and special characters, converting text to lowercase, and tokenizing the text into individual words or phrases.

Text Classification: Classify the security text data into different categories or labels based on their content. This can be done using supervised machine learning algorithms, where labeled training data is used to train a classifier to automatically assign categories to new, unlabeled data. This helps in organizing and structuring the text data for further analysis.

Named Entity Recognition (NER): Identify and extract named entities from the security text data, such as organization names, IP addresses, URLs, malware names, or vulnerability identifiers. NER techniques, often based on machine learning algorithms, can automatically recognize and tag these entities, enabling better understanding and analysis of the text data.

Sentiment Analysis: Determine the sentiment or opinion expressed in the security text data. Sentiment analysis techniques can help identify positive, negative, or neutral sentiments associated with specific security incidents, products, or services. This can provide insights into user experiences, customer satisfaction, or potential vulnerabilities.

Topic Modeling: Discover latent topics or themes within the security text data. Topic modeling algorithms, such as Latent Dirichlet Allocation (LDA), can automatically identify and extract underlying topics from a collection of documents. This helps in understanding the main subjects or issues discussed in the security text data.

Text Clustering: Group similar security text data together based on their content or characteristics. Clustering algorithms can identify patterns and similarities in the text data, enabling the identification of common themes, trends, or clusters of related incidents or vulnerabilities.

Text Summarization: Generate concise summaries or abstracts of the security text data. Text summarization techniques can automatically extract key information and important details from large volumes of text, providing a quick overview or summary of the content.

Visualization: Visualize the analyzed security text data to gain insights and communicate findings effectively. Techniques such as word clouds, bar charts, network graphs, or heatmaps can help visualize the frequency of terms, relationships between entities, or patterns in the text data.

Natural Language Processing (NLP) Techniques: Utilize various NLP techniques, such as named entity recognition, part-of-speech tagging, dependency parsing, or sentiment analysis, to extract deeper insights from the security text data. These techniques enable a more detailed understanding of the text content and its context.

By applying these techniques, organizations can gain valuable insights from security text data, such as identifying emerging threats, understanding user sentiments, detecting patterns of attacks, or improving incident response processes. These insights can inform decision-making, enhance security strategies, and improve overall cybersecurity defenses.

9.4.2 AI Applications in Security Communication

AI applications have significantly impacted security communication across various domains. AI algorithms are employed to detect and analyze security threats within communication networks. They can identify patterns indicative of malicious activities, such as abnormal traffic behavior, potential cyber attacks, or intrusion attempts. I-powered NLP tools monitor communication channels, including emails, chats, or social media, to detect suspicious content, phishing attempts, or language indicating potential security breaches. AI-driven behavioral analysis helps in recognizing anomalies in user behavior within communication systems. By understanding normal patterns of communication and interaction, AI can flag unusual behaviors that might indicate a security risk, such as unauthorized access or data leaks. AI contributes to enhancing encryption techniques and data security protocols, ensuring secure transmission and storage of sensitive information during communication processes. AI-based authentication systems use biometrics, behavioral analysis, or facial recognition to strengthen access controls, minimizing unauthorized access to communication channels or sensitive data [11]. AI models predict potential security threats by analyzing historical data and current trends. This proactive approach helps in preventing security breaches before they occur by implementing preemptive measures. AI-powered systems facilitate quick and automated responses to security incidents. They can isolate compromised systems, mitigate threats, and initiate incident response protocols, reducing the time to address security issues. AI aids in identifying and patching vulnerabilities in communication systems by scanning networks, applications, and devices for potential weaknesses that could be exploited by attackers.

AI's integration into security communication processes significantly improves the speed, accuracy, and efficiency of threat detection, response, and prevention. However, continuous adaptation and updates are essential to counter evolving security threats in today's dynamic digital landscape.

9.5 MACHINE LEARNING FOR MALWARE DETECTION

9.5.1 Signature-Based and Behavior-Based Approaches

Signature-based detection relies on known patterns or signatures of known threats to identify and block malicious activity. It involves comparing incoming data or files against a database of pre-defined signatures. If a match is found, the system can take appropriate action, such as blocking or quarantining the identified threat. Signature-based approaches are highly effective in detecting and blocking known malware, viruses, or other types of threats for which signatures are available. Since the signatures are specific to known threats, the false positive rate is generally low. Signature-based approaches cannot detect or prevent new or unknown threats that do not have a pre-defined signature. This makes them vulnerable to zero-day attacks or polymorphic malware. The signature database needs to be regularly updated to include new signatures for emerging threats. This requires continuous monitoring and maintenance.

Behavior-based detection focuses on analyzing the behavior of systems, networks, or users to identify abnormal or suspicious activities that may indicate a security threat [12]. It involves establishing a baseline of normal behavior and then monitoring for deviations from that baseline. Behavior-based approaches can detect and prevent unknown or zero-day threats that do not have pre-defined signatures. They can identify anomalies or deviations from normal behavior that may indicate a new or emerging threat. Behavior-based approaches can adapt to new attack techniques and evolve with changing threat landscapes. They can detect novel attack patterns or variations of known threats. Behavior-based approaches may generate more false positives compared to signature-based approaches. Legitimate activities that deviate from the established baseline can be flagged as suspicious, requiring additional analysis and investigation. Behavior-based approaches require continuous monitoring of systems, networks, and user activities to establish accurate baselines and detect anomalies. This can be resource-intensive and may require advanced analytics and machine learning techniques.

In practice, a combination of both signature-based and behavior-based approaches is often used to provide comprehensive threat detection and prevention. Signature-based methods are effective for known threats, while behavior-based methods help identify unknown or evolving threats. This layered approach enhances the overall security posture and reduces the risk of successful attacks [13].

9.5.2 Dynamic Threat Analysis

Dynamic threat analysis involves continuously monitoring, assessing, and responding to evolving security threats in real-time or near-real-time scenarios. This process is crucial in cybersecurity to detect and mitigate emerging threats that constantly evolve and change tactics to bypass traditional security measures. Here's how dynamic threat analysis works:

Systems continuously collect and analyze vast amounts of data from various sources, including network traffic, system logs, user behavior, and threat intelligence feeds. This real-time monitoring helps identify potential threats as they occur or even before they fully materialize. By understanding typical patterns of user behavior, system interactions, and network activities, dynamic threat analysis systems can identify deviations or anomalies that might indicate potential security threats. Behavioral analytics help in recognizing suspicious activities that don't conform to established norms [14] Utilizing machine learning and AI algorithms, dynamic threat analysis systems can adapt and learn from new data patterns and threats. These systems can detect sophisticated and previously unseen threats by continuously updating their models based on the latest information. Dynamic threat analysis incorporates threat intelligence from various sources, including global security databases, industry reports, and community-shared data. This integration allows for a broader understanding of emerging threats and helps in proactive threat detection. When potential threats are identified, automated responses can be initiated. These responses may include isolating affected systems, blocking suspicious IP addresses, or triggering alerts for human intervention. Automated responses help in mitigating threats swiftly to minimize potential damage. Dynamic threat analysis constantly evaluates and adjusts security measures based on the changing threat landscape. It involves updating security policies, patching vulnerabilities, and strengthening defenses to stay ahead of evolving threats. In the event of a security incident, dynamic threat analysis facilitates rapid incident response and forensic investigation. This allows organizations to understand the nature of the attack, its impact, and take appropriate measures to prevent future occurrences.

Dynamic threat analysis is a crucial component of modern cybersecurity strategies, providing a proactive and adaptive approach to combating constantly evolving threats in today's interconnected digital environments.

9.6 PREDICTIVE ANALYTICS FOR SECURITY INCIDENT RESPONSE

Predictive analytics can play a crucial role in enhancing security incident response by leveraging historical data, patterns, and machine learning algorithms to predict and prevent future security incidents. Here's how predictive analytics can be applied in security incident response:

Predictive analytics can analyze large volumes of threat intelligence data, including indicators of compromise (IOCs), vulnerabilities, and attack patterns. By identifying patterns and trends, predictive analytics can help security teams anticipate potential threats and proactively implement preventive measures. Predictive analytics can establish baselines of normal behavior for systems, networks, and users. By continuously monitoring and analyzing data, any deviations or anomalies from the established baseline can be detected. This helps in identifying potential security incidents or malicious activities in real-time. Predictive analytics can analyze user behavior, such as login patterns, access privileges, and data usage, to identify abnormal or suspicious activities. By leveraging machine learning algorithms, UBA can detect insider threats, compromised accounts, or unauthorized access attempts, enabling early detection and response [15].

Predictive analytics can assess the severity of security incidents based on historical data and contextual information. By analyzing past incidents and their impact, predictive models can help prioritize incident response efforts and allocate resources effectively. Predictive analytics can enable automated response and remediation actions based on pre-defined rules or machine learning models. For example, if a predictive model identifies a potential threat, it can trigger automated actions such as isolating affected systems, blocking suspicious IP addresses, or quarantining malicious files. Predictive analytics can assist in incident response planning by analyzing historical incident data and identifying common attack vectors, vulnerabilities, or weaknesses. This helps in developing proactive strategies, implementing preventive controls, and improving incident response processes. Predictive analytics can aid in proactive threat hunting by identifying potential indicators of compromise or emerging threats. By analyzing various data sources, such as logs, network traffic, or threat intelligence feeds, predictive models can help security teams identify hidden threats or patterns that may go unnoticed by traditional security controls. Predictive analytics can provide insights into the effectiveness of security controls, incident response processes, and overall security posture. By analyzing incident data and response metrics, organizations can identify areas for improvement, refine their security strategies, and enhance their incident response capabilities [16].

It's important to note that predictive analytics should be used in conjunction with other security measures, such as real-time monitoring, threat intelligence sharing, and human expertise. Predictive models should be regularly updated and refined to adapt to evolving threats and changing environments. By leveraging predictive analytics, organizations can proactively detect, respond to, and mitigate security incidents, reducing the impact of breaches and improving overall cybersecurity defenses.

9.7 AI IN ACCESS CONTROL AND AUTHENTICATION

9.7.1 BEHAVIORAL BIOMETRICS AND IDENTITY VERIFICATION

Behavioral biometrics in identity verification involves the use of unique behavioral patterns and characteristics exhibited by individuals to confirm their identities. Unlike traditional biometrics that rely on physical traits like fingerprints or iris scans, behavioral biometrics focus on how individuals interact with devices or systems. Analyzing the typing rhythm, speed, and patterns of an individual's keystrokes can create a unique biometric profile. It helps in confirming identities when users log in by comparing their typing behavior to the established profile. Behavioral biometrics consider the unique way individuals move and navigate a mouse or trackpad. Factors like speed, acceleration, and patterns of movement form a distinctive biometric profile for identity validation. While it falls within the realm of both behavioral and physiological biometrics, voice recognition analyzes individual speech patterns, tone, pitch, and speech dynamics to confirm identity during phone-based or voice-enabled authentication [17]. Analyzing gestures made on touchscreens or devices can form part of behavioral biometrics. Unique swipes, taps, or patterns traced by individuals can be used for identity verification. How individuals sign their names, including pressure, stroke sequence, and unique characteristics, can be used as behavioral biometrics for identity verification in documents or digital signatures. In some cases, the way an individual walks or moves can be captured as a behavioral biometric for identity verification. This method is particularly applicable in scenarios like surveillance or access control systems.

Behavioral biometrics add an extra layer of security to identity verification processes. As these traits are unique to individuals and difficult to replicate, they provide an additional level of confidence in confirming someone's identity. However, ensuring accuracy and reliability in capturing and analyzing these behavioral patterns is crucial for effective identity verification systems.

9.7.2 Multi-Factor Authentication with AI

Multi-factor authentication (MFA) is a security measure that combines multiple forms of authentication to verify the identity of users. By incorporating artificial intelligence (AI) techniques into MFA, behavioral biometrics can be used as one of the factors for identity verification (Figure 9.2). Here's how AI can enhance MFA with behavioral biometrics:

- **Continuous Authentication:** AI algorithms can continuously analyze and monitor user behavior patterns, such as keystrokes, mouse movements, or gestures, during a session. This allows for ongoing authentication and can detect anomalies or suspicious activities that may indicate unauthorized access.
- **Machine Learning Models:** AI-powered machine learning models can be trained to recognize and authenticate individual behavioral biometric patterns. These models can adapt and improve over time by learning from user interactions, making the authentication process more accurate and reliable.
- **Real-time Risk Assessment:** AI algorithms can assess the risk level associated with a user's behavioral biometrics in real-time. By comparing the current behavior to the established profile, AI can identify potential fraud attempts or compromised accounts and trigger additional security measures or alerts.
- **Contextual Analysis:** AI can analyze contextual information, such as device location, IP address, or time of access, along with behavioral biometrics. This helps in determining the legitimacy of the authentication request and adds an extra layer of security.
- **Adaptive Authentication:** AI can dynamically adjust the authentication requirements based on the risk level associated with a user's behavioral biometrics. For example, if a user's behavior deviates significantly from their established profile, AI can prompt for additional authentication factors or step-up authentication.
- **Fraud Detection:** AI algorithms can detect patterns and anomalies in behavioral biometrics that may indicate fraudulent activities, such as account takeover attempts or impersonation. By analyzing large volumes of data, AI can identify suspicious behavior and trigger appropriate actions to prevent fraud.

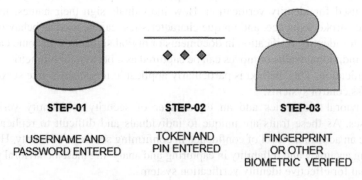

FIGURE 9.2 Multi-Factor Authentication.

Artificial Intelligence for Information System Security

User Experience Optimization: AI can analyze user behavior patterns to personalize the authentication experience. By understanding how individuals interact with devices or systems, AI can streamline the authentication process, reducing friction and enhancing user experience without compromising security.

It's important to note that while AI-enhanced MFA with behavioral biometrics provides an additional layer of security, it should be used in conjunction with other authentication factors, such as passwords, tokens, or biometrics. This multi-layered approach helps mitigate the risk of single-point failures and provides a robust authentication mechanism. Additionally, organizations should ensure the privacy and security of the behavioral biometric data collected, adhering to relevant regulations and best practices.

9.8 SECURING CLOUD ENVIRONMENTS WITH AI

Securing cloud environments is a critical aspect of maintaining data privacy and protecting against cyber threats. Artificial intelligence (AI) can play a significant role in enhancing cloud security by leveraging its capabilities in data analysis, anomaly detection, and threat intelligence [18]. AI-powered security solutions can analyze large volumes of data generated in cloud environments, including logs, network traffic, and user behavior, to detect and prevent threats. A framework for secure cloud computing environments is shown in Figure 9.3.

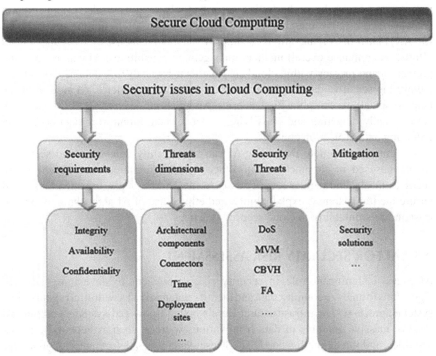

FIGURE 9.3 A framework for secure cloud computing environments.

Machine learning algorithms can identify patterns and anomalies that may indicate malicious activities, such as unauthorized access attempts, data exfiltration or malware infections. I can analyze user behavior within cloud environments to establish baselines of normal behavior. By continuously monitoring and analyzing user activities, AI can detect deviations from the established baseline, flagging suspicious behavior that may indicate insider threats or compromised accounts. AI algorithms can identify anomalies in cloud environments by comparing current activities to historical data and established patterns. This helps in detecting unusual network traffic, abnormal resource usage, or unauthorized configuration changes that may indicate a security breach [19].

AI can automate security processes in cloud environments, such as vulnerability scanning, patch management, or incident response. By leveraging AI algorithms, organizations can streamline security operations, reduce response times, and improve overall incident management. AI can analyze threat intelligence feeds, security blogs, and other sources of information to identify emerging threats and vulnerabilities. By continuously monitoring and analyzing this data, AI can provide real-time insights and proactive recommendations to secure cloud environments against new and evolving threats. User and Entity Behavior Analytics (UEBA) AI-powered UEBA solutions can analyze user and entity behavior within cloud environments to detect suspicious activities or unauthorized access attempts. By correlating various data sources, such as logins, file access, or resource usage, AI can identify potential security incidents and trigger appropriate response actions. AI can automate incident response processes by orchestrating security tools, analyzing alerts, and initiating response actions. By integrating AI-powered security orchestration and response (SOAR) platforms, organizations can improve incident response times, reduce manual efforts, and enhance overall incident management capabilities. AI can assist in data protection and privacy within cloud environments by analyzing data access patterns, identifying sensitive data, and enforcing data protection policies. AI algorithms can help organizations comply with data privacy regulations, such as GDPR or CCPA, by automatically detecting and classifying sensitive data, monitoring data access, and enforcing data protection controls.

It's important to note that while AI can significantly enhance cloud security, it should be used in conjunction with other security measures, such as encryption, access controls, and regular security assessments. Additionally, organizations should ensure the transparency, explainability, and ethical use of AI algorithms to maintain trust and accountability in cloud security practices.

9.9 AI FOR SECURITY RISK ASSESSMENT

AI plays a significant role in security risk assessment by leveraging advanced algorithms to analyze complex data and identify potential risks within an organization's systems, processes, or infrastructure. Here's how AI-powered systems continuously analyze massive amounts of data from various sources, such as network logs, user activities, and threat intelligence feeds. These systems can detect patterns, anomalies, and indicators of potential security threats, helping to predict and proactively mitigate risks [20]. AI algorithms scan and assess systems for vulnerabilities by analyzing configurations, code, or network traffic. They can identify potential weaknesses that could be exploited by attackers, enabling organizations to patch or mitigate these vulnerabilities before they are exploited. AI-based systems monitor user behavior and interactions within networks and systems. They can detect deviations from normal behavior, such as unauthorized access attempts or unusual data access patterns, indicating potential security risks or insider threats.

AI automates parts of the risk assessment process by analyzing data at scale and providing risk scores or prioritizing threats based on severity. This helps security teams focus on the most critical issues that require immediate attention. AI enables adaptive security measures by learning from past incidents and continuously improving threat detection capabilities. These systems adapt and evolve to counter new and evolving threats, making security risk assessments more effective over time. AI assists in incident response by providing real-time analysis and insights during security incidents. It helps in understanding the nature and scope of an attack, enabling faster and more effective responses to mitigate risks. Using historical data and machine learning models, AI can predict future security risks based on trends and patterns. This information allows organizations to proactively implement mitigation strategies and strengthen their security posture.

AI-driven security risk assessment enhances the accuracy, speed, and efficiency of identifying and mitigating potential threats and vulnerabilities. However, it's essential to ensure proper training of AI models, data privacy compliance, and human oversight to avoid biases and false positives in risk assessment processes.

9.10 CHALLENGES AND LIMITATIONS OF AI IN INFORMATION SECURITY

AI systems can be susceptible to adversarial attacks where malicious actors intentionally manipulate or deceive the AI models. They exploit vulnerabilities by inputting specially crafted data to deceive the system, causing it to make incorrect decisions or predictions. AI models can inherit biases from the data they are trained on, leading to biased decisions or outcomes. This bias can disproportionately impact certain groups or make the system less effective in diverse environments, especially in applications such as hiring or risk assessment.

AI in security often requires access to sensitive data for analysis. Safeguarding this data from breaches or unauthorized access is crucial. Balancing the need for data access with privacy regulations and ensuring secure data handling remains a challenge. Complex AI algorithms, particularly deep learning models, can lack transparency or interpretability. Understanding why AI systems make specific decisions or predictions is challenging, which can be a limitation in critical security applications where explanations are required. Cyber threats evolve rapidly, and attackers adapt their methods. AI-based security measures must keep pace with these changes. If AI models are static or not updated regularly, they might become less effective against emerging threats.

Implementing and maintaining AI-driven security systems often require significant computational resources and expertise. Small or resource-constrained organizations might find it challenging to adopt and manage AI-based security solutions. Relying solely on AI for security can create a false sense of security. Human expertise and oversight remain essential to complement AI systems, as humans can provide context, intuition, and critical thinking that AI might lack. Addressing these challenges involves ongoing research and development in AI ethics, enhancing model interpretability, robustness testing against adversarial attacks, improving data quality, and developing AI technologies that are adaptable to evolving threats while ensuring compliance with privacy and regulatory standards

9.11 FUTURE TRENDS IN AI FOR INFORMATION SYSTEM SECURITY

Several future trends are expected to shape the landscape of AI in information system security:

Explainable AI (XAI): Enhancing the interpretability of AI models is crucial. Future AI systems will focus on providing explanations for their decisions, allowing security professionals to understand and trust the reasoning behind the system's actions.

AI-Powered Autonomous Security Systems: Advancements in AI will lead to the development of autonomous security systems capable of detecting, analyzing, and responding to threats in real-time without human intervention. These systems will adapt and learn from new threats independently.

Federated Learning for Security: Federated learning, where models are trained across multiple decentralized devices or servers, will play a significant role in maintaining privacy and security. It allows AI models to be trained collaboratively without sharing sensitive data.

AI for Behavioral Biometrics: Behavioral biometrics, leveraging AI, will expand beyond typing patterns or mouse movements to include more sophisticated behavioral traits for identity verification, such as behavioral analysis through IoT devices or user interaction patterns.

Privacy-Preserving AI: There will be a focus on developing AI models that can perform computations while preserving the privacy of sensitive data. Techniques like homomorphic encryption and secure multi-party computation will enable secure analysis of encrypted data.

AI-Enabled Threat Hunting: AI will assist security analysts by proactively hunting for potential threats, identifying hidden patterns, and predicting future attack vectors. This proactive approach will help in countering sophisticated cyber threats.

AI-driven Cyber Range Simulations: Simulations powered by AI will be used to mimic cyberattacks, enabling organizations to test and strengthen their security measures in a controlled environment. This proactive testing will enhance readiness against evolving threats.

Ethical AI Governance Frameworks: As AI becomes more embedded in security, there will be increased emphasis on establishing ethical guidelines and governance frameworks to ensure responsible and ethical use of AI in security applications.

These trends represent the evolving landscape where AI will continue to be a driving force in strengthening information system security. Embracing these advancements while addressing associated challenges will be key in securing digital infrastructures in the future.

9.12 CONCLUSION: THE FUTURE LANDSCAPE OF AI IN INFORMATION SYSTEMS

In conclusion, the future landscape of AI in information systems is poised for transformative advancements with significant implications for security, efficiency, and innovation. AI's role will extend beyond its current applications, shaping various aspects of information systems. The collaborative efforts of researchers, developers, policymakers, and stakeholders will play a crucial role in shaping a future where AI contributes to safer, more efficient, and ethically governed information systems. Ultimately, AI's evolution will revolutionize how we interact with, secure, and derive insights from information systems, propelling us toward a more intelligent and resilient digital future.

REFERENCES

1. Ren, Q., 2021. RETRACTED: Application Analysis of Artificial Intelligence Technology in Computer Information Security. *Journal of Physics: Conference Series*, 1744. p. 042221. https://doi.org/10.1088/1742-6596/1744/4/042221.
2. Grusho, A., Grusho, N., Zabezhailo, M., Timonina, E., 2016. Intelligent Data Analysis in Information Security. *Automatic Control and Computer Sciences*, 50, pp. 722–725. https://doi.org/10.3103/S0146411616080307.
3. He, S., Shi, X., Huang, Y., Chen, G., Tang, H., 2022. Design of Information System Security Evaluation Management System based on Artificial Intelligence. *2022 IEEE 2nd International Conference on Electronic Technology, Communication and Information (ICETCI)*, pp. 967–970. https://doi.org/10.1109/icetci55101.2022.9832131.
4. Purchina, ., Poluyan, A., Fugarov, D., 2023. An Algorithm based on Artificial Intelligence for Solving Information Security tasks. *E3S Web of Conferences*. https://doi.org/10.1051/e3sconf/202337103066.
5. Pavlova, G., Tsochev, G., Yoshinov, R., Trifonov, R., Manolov, S., 2017. Increasing the Level of Network and Information Security Using Artificial Intelligence. *Fifth International Conference on Advances in Computing, Communication and Information Technology*, pp. 83–88. https://doi.org/10.15224/978-1-63248-131-3-25.
6. Antipov, S., Vagin, V., Morosin, O., Fomina, M., 2018. Protection of Information in Networks Based on Methods of Machine Learning, *Artificial Intelligence: 16th Russian Conference, RCAI 2018*, pp. 273–279.

7. Yuan, Q., Tan, X., 2021. Research on Application of Artificial Intelligence in Network Security Defence. *Journal of Physics: Conference Series*, 2033, p. 012149. https://doi.org/10.1088/1742-6596/2033/1/012149.
8. Moon, J., Kim, S., Song, J., Kim, K., 2021. Study on Machine Learning Techniques for Malware Classification and Detection. *KSII Transactions on Internet and Information Systems*, 15, pp. 4308–4325. https://doi.org/10.3837/tiis.2021.12.003.
9. Xiao-Yang, Y., 2011. Study on Development of Information Security and Artificial Intelligence. *2011 Fourth International Conference on Intelligent Computation Technology and Automation*, vol. 1, pp. 248–250. https://doi.org/10.1109/ICICTA.2011.72.
10. Gang, Z., 2012. Application of Artificial Intelligence to Information System Security Risk Assessment. *Journal of Beijing Information Science Technology University*, Vol. 6, p. 45.
11. Kokoshin, A., 2019. Artificial Intelligence and Some Issues of Russian Security Provision. *естник Российской академии наук*, 89, 437–439. https://doi.org/10.31857/S0869-5873895437-439.
12. Kovtsur, M., Mikhailova, A., Potemkin, P., Ushakov, I., Krasov, A., 2020. Guidelines for Using Machine Learning Technology to Ensure Information Security. *2020 12th International Congress on Ultra Modern Telecommunications and Control Systems and Workshops (ICUMT)*, pp. 285–290. https://doi.org/10.1109/ICUMT51630.2020.9222417.
13. Zhou, Y., Ouyang, M., Shu, S., 2021. Artificial Intelligence Student Management based on Embedded System. *Microprocessors and Microsystems*, 83, pp. 103976. https://doi.org/10.1016/J.MICPRO.2021.103976.
14. Wazid, M., Das, A., Park, Y., 2021. Blockchain-Envisioned Secure Authentication Approach in AIoT: Applications, Challenges, and Future Research. *Wireless Communications and Mobile Computing*, 2021, p. 3866006. https://doi.org/10.1155/2021/3866006.
15. Kumar, M., Patil, H., 2022. Challenges and Solution of Artificial Intelligence in Cyber Security. *Recent Trends in Artificial Intelligence its Applications*. Vol. 8, p. 15. https://doi.org/10.46610/rtaia.2022.v01i02.006.
16. Garbuk, S., 2021. Tasks of Technical Regulation of Intelligent Information Security Systems, Vol. 10, p. 89. https://doi.org/10.21681/2311-3456-2021-3-68-83.
17. Pieters, W., 2011. Explanation and Trust: What to Tell the User in Security and AI? *Ethics and Information Technology*, 13, pp. 53–64. https://doi.org/10.1007/s10676-010-9253-3.
18. Erokhin, S., Zhuravlev, A., 2020. A Comparative Analysis of Public Cyber Security Datasets. *2020 Systems of Signal Synchronization, Generating and Processing in Telecommunications (SYNCHROINFO)*, pp. 1–7. https://doi.org/10.1109/SYNCHROINFO49631.2020.9166001.
19. Zhong, B., 2020. Research on Computer Network Security Analysis Modeling Based on Artificial Intelligence Technology. *Proceedings of the 2020 International Conference on Aviation Safety and Information Technology*. https://doi.org/10.1145/3434581.3434709.
20. Parati, N., Malik, L., Joshi, A., 2008. Artificial Intelligence Based Threat Prevention and Sensing Engine: Architecture and Design Issues. *2008 First International Conference on Emerging Trends in Engineering and Technology*, pp. 304–307. https://doi.org/10.1109/ICETET.2008.52.

10 Optimized Image Recognition for Smart Information Generations

A Comprehensive Evaluation of Image Processing Techniques

Gurmeet Kaur Saini, Aleem Ali, and Nawaf R. Alharbe

10.1 INTRODUCTION TO PROPOSED MODEL DESIGN

10.1.1 Speeded Up Robust Features (SURF)

Speeded Up Robust Features (SURF) is a local feature detector and descriptor that is helpful for 3D reconstruction, object recognition, registration, and classification. It was somewhat inspired by the scale-invariant feature transform (SIFT) descriptor. The normal version of SURF, according to its creators, is faster by several times and more resistant to different picture modifications than SIFT. To find interest spots, SURF uses an integer approximation of the determinant of the Hessian blob detector, which may be found with three integer operations using a recomputed integral image. Its feature descriptor is derived from the sum of the Haar-wavelet response around the point of interest [37]. These can also be computed using the integral image. SURF descriptors are useful for generating three-dimensional scenes, tracking objects, locating and identifying people or objects, and extracting points of interest [2].

SURF serves as a point-of-interest detector and descriptor when an image is converted to coordinates using the multi-resolution pyramid approach. This allows one to duplicate the original image with a Pyramidal Gaussian or Laplacian Pyramid shape, resulting in an image with the same size but less bandwidth [38]. This produces a unique blurring of the original image that is only possible in Scale-Space. The scale invariance of the points of interest is guaranteed by this method. When the multi-resolution pyramid technique is used to transform an image to coordinates, SURF functions as a point-of-interest detector and descriptor. This enables the production of an image that is the same size but has less bandwidth by copying the original with a Pyramidal Gaussian or Laplacian Pyramid shape [38]. In doing so, the original image

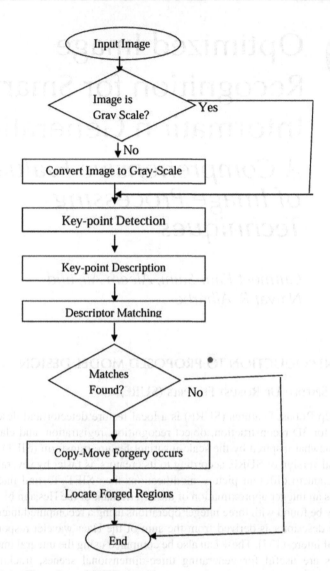

FIGURE 10.1 Flow chart of SURF algorithm.

is blurred in a way that is exclusive to Scale-Space [3] (Figure 10.1). This technique ensures the scale invariance of the points of interest.

The following actions are taken to identify copy-move forgery:

First, use the forged image.
Step 2: Verify that the provided image is in grayscale before performing any more processing. If it isn't, convert the picture to grayscale first.
Step 3: The feature extraction and description vectors are then completed using the SURF technique.
Step 4: Next, matching is carried out to find the forged portion in the digital picture.

Optimized Image Recognition for Smart Information Systems

Algorithm 1: SURF Algorithm

- Open the original picture.
- Obtain the original image's size.
- In the event that the initial image depth is 3.
- Transform the picture to grayscale.
- Overlay the original image with SURF.
- Utilize the SURF function to obtain the SURF data matrix.
- Open the alleged picture.
- Determine the suspected image's size.
- In the event that the suspected image depth is 3.
- Transform the picture to grayscale.
- Over the suspected image, apply SURF.
- Utilize the SURF function to obtain the SURFdata matrix.
- Apply the k-nearest neighbor (kNN) technique to assess the point locations between the point location arrays derived from the original and suspected image.
- The kNN technique is used to determine the Euclidean distance.
- Analyze the point descriptors at the matching position.
- Give back the points that match.
- Give back the original image's non-matching points (Points of Removal).

10.1.2 SCALE-INVARIANT FEATURE TRANSFORM (SIFT)

Scale-Invariant Feature Transform, or SIFT, is a technique used in computer vision to locate and describe local features in images. Applications include robotic mapping and navigation, match moving, gesture recognition, object recognition, image stitching, 3D modelling, video tracking, and individual wildlife identification [4].

Any object in an image can have an interesting point extracted from it to provide a "feature description" of the object. This description, which was extracted from a training image, can be used to help identify an object when searching through a test image that contains a lot of other things. To achieve accurate recognition, it is essential that the characteristics extracted from the training image can still be identified in the presence of noise, changes in illumination, and scale variations. These points are usually observed on highly contrasted portions of the image, such as object edges.

First, an item's SIFT key points are extracted from a set of reference pictures and entered into a database. Each feature in a new image is compared individually to this database to identify objects, and candidate matching features are found by calculating the Euclidean distance between their feature vectors [5].

Finding subsets of important points that agree on the location, scale, and orientation of the item in the new image allows one to remove good matches from the whole collection of matches. To find consistent clusters quickly, an effective hash table implementation of the generalized Hough transform is used. Each cluster consisting of three or more features that agree on an object's pose is removed after a more comprehensive model verification.

Lastly, the likelihood that a certain set of attributes indicates the existence of an object is ascertained, taking into account the accuracy of fit and the quantity of possible false matches.

Figure 10.2 illustrates how the SIFT algorithm functions.

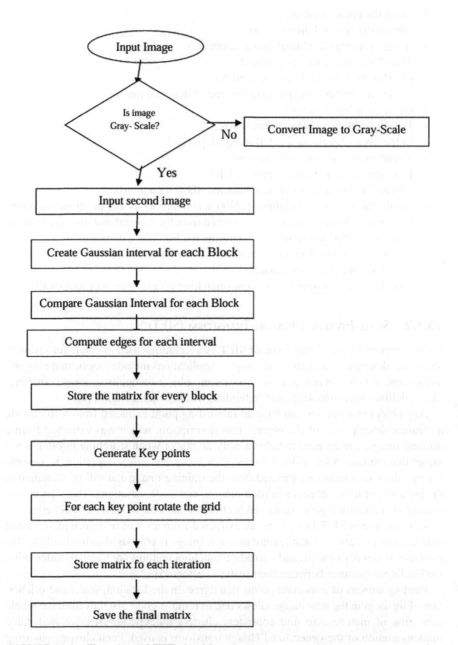

FIGURE 10.2 Flow chart of SIFT algorithm.

Optimized Image Recognition for Smart Information Systems

Algorithm 2: SIFT Algorithm

Second Algorithm: SIFT Method

- Open the picture.
- Image resampling to twice the size every octave.
- Make intervals of Gaussian blur.
- Construct Gaussian intervals with differences.
- For every interval, calculate the edges.
- For every important point
    a) Turn the sample grid to align the essential points.
    b) Select an area and produce a description.
    c) Conclude for
- Preserve pyramid pictures if desired.
- Store the resultant pictures.
- Preserve descriptors.

10.1.3 COMBINED WORK OF SIFT AND SURF

To demonstrate a new approach to copy-move forgery detection, the image will be first transformed into the wavelet domain using DWT, and the features will then be extracted from the modified image using SIFT. The second level of feature transformation will employ SURF [6]. As wavelet produces multispectral components, features become more common. After obtaining the interest point feature descriptor, we will try to find matching between these feature descriptors to assess whether or not post-processing manipulation of the given image has taken place. Our findings show that the optimal choice is to combine the powerful performance and superior computing efficiency of SURF and SIFT features [7].

Algorithm 3: Proposed Algorithm

1. Obtain the SIFT Algorithm's non-matching critical points.
2. Find the SURF Algorithm non-matching crucial points.
3. Use an SVM classifier to eliminate critical spots that have identical pixel values.
4. Put every important point in a matrix.
5. Label the image's important areas.

The working of the proposed system is shown in Figure 10.3.

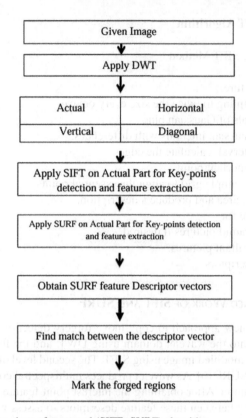

FIGURE 10.3 Flow chart of proposed (SIFT+SURF) algorithm.

10.2 SIMULATION TOOL DESCRIPTION

MathWorks created the scientific computer environment known as MATLAB, which stands for Matrix-Laboratory. The main tasks it performs include matrix manipulation, function plotting, algorithm implementation, user interface creation, etc. For calculations that necessitate the heavy use of arrays and graphical data analysis, this makes it perfect [8].

A robust program can be developed in only a few lines of code thanks to the way the MATLAB programming language is designed. Unlike traditional general-purpose programming languages like C++ or Java, it can solve complicated problems with a comparatively small number of statements. It is widely acknowledged in companies and research related to science, economics, engineering, and other fields because of its wide range of applications [9].

10.3 VERSION USED AND SYSTEM SPECIFICATIONS

- The version used for MATLAB is 2015 a.
- Intel core 2 DUO processor is required.

Optimized Image Recognition for Smart Information Systems

- 2 GB RAM and 4-5 GB Disk Space is required.
- The platform used is windows 7(64-bit).

10.4 RESULT AND ANALYSIS

The following systems were created with MATLAB 2015 and tested on an Intel Core i3 with 4GB RAM running Windows 7. This platform should be considered the least hardware necessary because the algorithms for key-point removal & injection and picture forgery detection may have been modified for greater accuracy on a more powerful testing platform. These days, automated image forgery detection is the ultimate goal of artificial intelligence in computer vision. This is undoubtedly the most challenging and ambitious computer vision topic now being researched, and it's not just a fascinating theoretical problem either—the real world genuinely needs a system like this [10]. Some of the computational results of the programme we have proposed are presented in this portion of the thesis. The details in Experimental Results 1 and 2 are similar between the test image and the equivalent image stored in the database, notwithstanding certain fictitious pixel groupings (Figure 10.4). The test image and the comparable image were not created in the same way, as Experimental Results 3 and 4 show.

Approximately 100 images of various picture forgeries were collected in order to evaluate the aforementioned technologies. The fully automated image forgery detection system, the automated image forgery detection system, the manual image forgery detection system, and the fully automated image forgery detection and detection

FIGURE 10.4 Testing Images from the proposed image forgery detection model.

FIGURE 10.5 The image forgery image from the dataset.

system are all tested using several frontal view photos per test subject [38]. The first image was taken on a white background in 'great' lighting. This would be used by the picture fraud detection system as the acknowledged frontal view hand image. The researcher assigned the image's ambient state a grade of "A." The other frontal view pictures were taken in lighting conditions that were getting worse and sometimes with an entirely dark background. These would be used as test photos by the frontal view picture forgery detection system. We tried everything to change the lighting in the area where the photos were taken in order to assess how reliable the systems were. The environmental status of the photograph was classified as 'B'. To help with pose invariant image forgery detection, the following data was collected. Nine known images were collected from each participant, and three (unknown) pictures of each subject were taken while they were posed at angles between the nine known images [11]. For the nine images that are available, the test subject was photographed in the positions that are depicted in Figures 10.5. Since frontal view image fraud identification is more reliable than pose invariant image forgery detection, the data was gathered under very strict controls. The automatic exposure of the digital camera used to produce the faked photographs had a significant negative impact on several of the test subjects' images, forcing them to be discarded.

10.4.1 Controlled Database of Image Forgeries

Both controlled and uncontrolled datasets were obtained. Because the controlled dataset was gathered against a similar background, it can produce the highest degree of accuracy. It has been demonstrated that the database that was compiled under ideal

Optimized Image Recognition for Smart Information Systems

FIGURE 10.6 The controlled image forged obtained using the 13 mega pixel real-aperture camera.

conditions is the most accurate. The various items (people) have been added to the database under regulation [12]. The purpose of getting the hand images was to make it easier for people to engage with computers and the operating room robots, which need to understand hand language in order to function. With the help of our work, medical personnel can now remotely control robotic hands to improve operative accuracy. A few screenshots from the dataset are shown below: Both controlled and uncontrolled datasets were obtained. Because the controlled dataset was gathered against a similar background, it can produce the highest degree of accuracy [14, 28, 26]. It has been demonstrated that the database that was compiled under ideal conditions is the most accurate. The various items (people) have been added to the database under regulation. The purpose of getting the hand images was to make it easier for people to engage with computers and the operating room robots, which need to understand hand language in order to function. With the help of our work, medical personnel can now remotely control robotic hands to improve operative accuracy. A few screenshots from the dataset are shown in Figure 10.6.

10.4.2 Image Forgery Dataset (Natural Hand Images)

Additionally, the database was gathered from online sources by compiling royalty-free picture forgery [39]. This database has been used to test the suggested model's adaptation and flexibility in scenarios other than the controlled ones. The possible picture forgery locations suggested by the fully automated image forgery detection system are here verified by correlation with the average image forgery. Images for Conditions A and B are the same as in the prior test. Images of the image forgery in vivid colors and with enough light are included in condition A. Images with low brightness and normal or less than normal light conditions are subject to Condition B. Overall tested condition of images: successful detection failures [41].

10.4.3 Performance Evaluation Measures

Important terms that are required to comprehend the performance measurements include the following:

The number of altered photos that are labelled as tampered is known as TP (True Positive).
False Negative, or FN, is the quantity of altered photos that are accepted as real.
The quantity of genuine photos that are deemed to be real is known as TN (True Negative).
False Positive, or FP, is the quantity of real photos that are labelled as altered.

The words "true positives," "true negatives," "false positives," and "false negatives" refer to the comparison of test classifier results with reliable outside assessments in classification tasks. Positive and negative denote the classifier's prediction, whereas true and false denote whether or not the prediction agrees with the outside assessment [13, 15, 27].

10.4.4 Performance Evaluation Parameters

10.4.4.1 Accuracy

The percentage of photos that the classifier properly classifies is known as accuracy. It is calculated as follows:

$$Accuracy = (TP + TN) / (TP + TN + FN + FP)$$

Table 10.1 shows the results of testing 100 photos, of which 95 successfully identified (also known as True positive) images and 5 failures. The values of false positives are three, false negatives are one, and true negatives are one. The graph in Figure 10.7 displays the overall accuracy:

The graph in Figure 10.7 displays the genuine positive value, or 95, which indicates that the suggested model was successfully detected. There are a total of five failures, and the accuracy of the proposed model is 96%.

There are a total of 100 photographs evaluated in Table 10.2, 50 of which are color images and the remaining 50 are low brightness images. Out of the 48 images that are properly recognized for Bright Color, 2 have been incorrectly identified as forged, and 1 image has been accurately identified as non-forged. 47 low brightness

TABLE 10.1
shows evaluation measures for accuracy

Condition of Image	Total images tested	True positive (TP)	True negative (TN)	False positive (FP)	False negative (FN)	Accuracy (in %)
Overall	100	95	1	3	1	96.00%

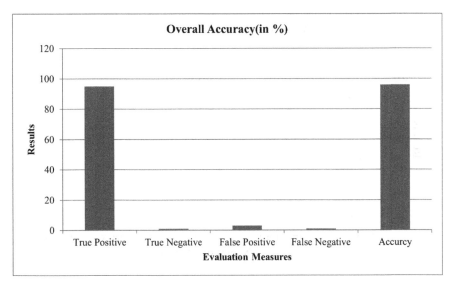

FIGURE 10.7 shows overall accuracy of proposed model.

TABLE 10.2
shows accuracy for Bright Color Images and Low Brightness Images

Condition of Image	Total tested	True positive	True negative	False positive	False negative	Accuracy (in %)
Bright Color Images	50	48	1	1	0	98.00 %
Low Brightness Images	50	47	0	2	1	94.00 %

photos are successfully discovered; 2 wrongly detected as forged images, 2 correctly detected as non-forged images, and 1 falsely missed but forged image are among the false positives. Figure 10.8 is the accuracy graph for each of the two conditions [16, 17, 18, 29, 25, 32].

The graph in Figure 10.8 displays the suggested model's accuracy for photos with vibrant colors. The graph displays the maximum accuracy of 98% that has been recorded.

This graph in Figure 10.9 displays the suggested model's accuracy for photos with low brightness. The graph illustrates the greatest accuracy, which is 94% recorded.

With the use of evaluation metrics, this graph in Figure 10.10 compares the correctness of the suggested model for images with bright colors and low brightness. The graph displays the maximum accuracy, which is 98% for bright color photographs and 94% for low brightness images [19, 31, 23, 24].

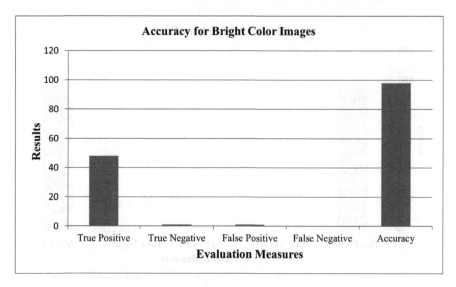

FIGURE 10.8 shows overall accuracy for Bright Color Images.

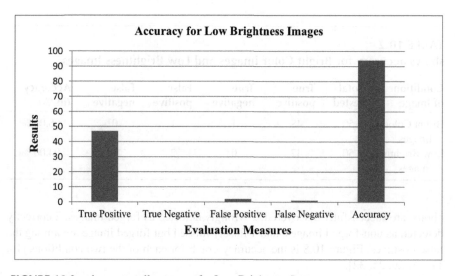

FIGURE 10.9 shows overall accuracy for Low Brightness Images.

10.4.4.2 Precision:

Precision, which is often referred to as positive predictive value, is the percentage of relevant instances recovered. It is predicated on a relevance metric. It is computed as follows: Precision = TP / TP + FP

The accuracy of the recommended model is seen in Table 10.3 for images with vivid colors. The maximum accuracy of 98% that has been recorded is shown in Table 10.3.

Optimized Image Recognition for Smart Information Systems

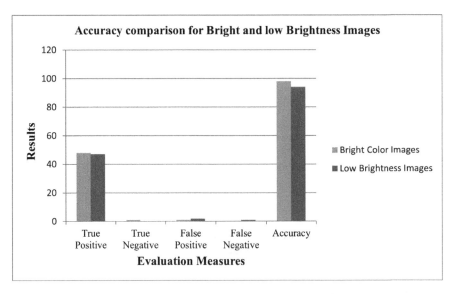

FIGURE 10.10 Accuracy comparison for Bright Color and Low Brightness Images.

TABLE 10.3
shows evaluation measures for Precision

Condition of Image	Total images tested	True Positive (TP)	True Negative (TN)	False Positive (FP)	False Negative (FN)	Precision (in %)
Overall	100	95	1	3	1	96.93%

Figure 10.11 displays the suggested model's total precision, or positive predictive value. The graph displays the greatest precision, which is recorded at 96.33%.

There are a total of 100 photographs evaluated in Table 10.4, 50 of which are color images and the remaining 50 are low brightness images. 48 photos of bright color are successfully identified, 1 image is accurately identified as not-forged, and 1 image is mistakenly identified as forged. Figure 10.12 displays the Bright Color Precision Graph [20, 21, 22, 26, 28, 30].

The precision value of the suggested model for bright color images is displayed in Figure 10.12. The graph displays the greatest precision rate, which is recorded at 99.95%.

Table 10.5 reveals that a total of 50 photographs were checked for low brightness; 47 of those images were properly recognized, 2 of those images were mistakenly identified as forged, 2 of those images were correctly identified as non-forged, and 1 image was mistakenly missed but turned out to be forged. Below is the Precision Graph for Low Brightness images:

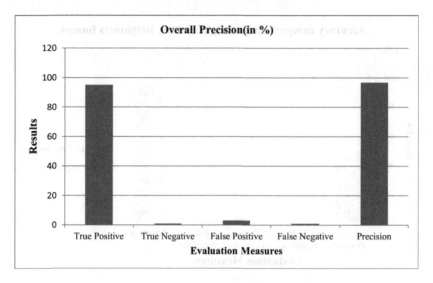

FIGURE 10.11 shows overall Precision value of Proposed Model.

TABLE 10.4
shows precision for Bright Color Images

PROPERTY	Bright Color Images
True Positive	48
True Negative	1
False Positive	1
False Negative	0
Precision (Positive Predictive Value)	97.95%

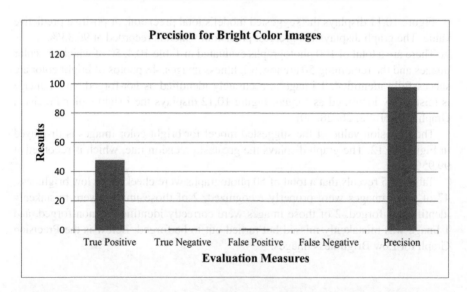

FIGURE 10.12 shows Precision value of Proposed Model for Bright Color Images.

Optimized Image Recognition for Smart Information Systems

TABLE 10.5
shows precision for Low Brightness Images

PROPERTY	Low Brightness Images
True Positive	47
True Negative	0
False Positive	2
False Negative	1
Precision (Positive Predictive Value)	95.91%

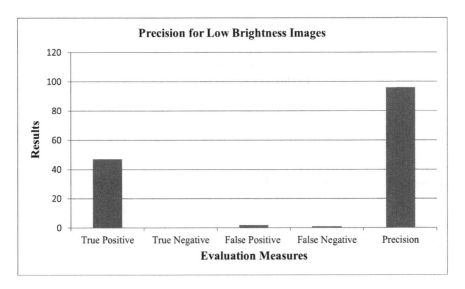

FIGURE 10.13 shows Precision value of Proposed Model for Low Brightness Images.

Figure 10.13 compares the recall of the proposed model for images with vivid colors and low brightness using assessment measures. The graph shows the highest recall, which is 100% for color images with high brightness and 97.11% for images with low brightness.

With the use of evaluation metrics, Figure 10.14 illustrates the accuracy comparison of the suggested model for images with bright colors and low brightness. The graph displays the greatest precision, which is reported at 97.95% for bright color images and 95.71% for low brightness images.

10.4.4.3 Recall

The percentage of pertinent instances that are retrieved is known as recall. It is predicated on a comprehension and assessment of significance. Another name for this is sensitivity. It is computed as follows:

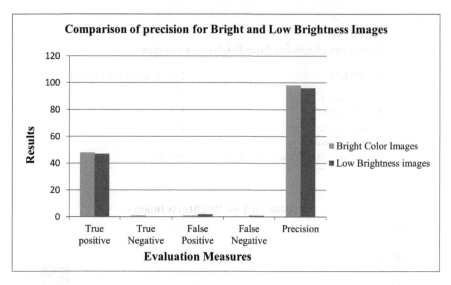

FIGURE 10.14 Precision comparison for Bright Color and Low Brightness Images.

TABLE 10.6
shows evaluation measures for Recall

Condition of Image	Total images tested	True Positive (TP)	True Negative (TN)	False Positive (FP)	False Negative (FN)	Recall (in %)
Overall	100	95	1	3	1	98%

$$Recall = TP / TP + FN$$

Table 10.6 demonstrates that a total of 100 photos were evaluated, of which 95 were found to be true positives—that is, correctly detected. There are a total of five failures: one true negative, one false negative, and three false positives. The graph in Figure 10.15 displays the total recall:

Figure 10.15 displays the proposed model's overall recall, sometimes referred to as sensitivity. The graph displays the greatest recall, which is recorded at 98%.

There are a total of 100 photographs evaluated in Table 10.7, 50 of which are color images and the remaining 50 are low brightness images. 48 photos of bright color are successfully identified, 1 image is accurately identified as not-forged, and 1 image is mistakenly identified as forged. Figure 10.16 is the Bright Color image recall graph.

The recall value of the suggested model for bright color images is displayed in Figure 10.16. The graph displays the maximum recall rate, which is recorded at 100%.

Optimized Image Recognition for Smart Information Systems 207

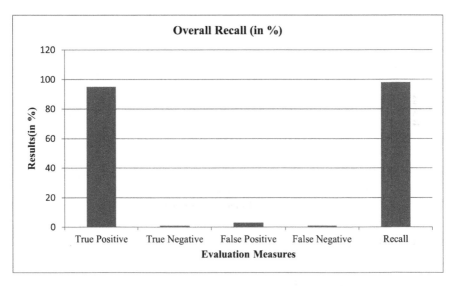

FIGURE 10.15 shows overall Recall value of Proposed Model.

TABLE 10.7
shows parameter values for Bright Color Images

PROPERTY	Bright Color Images
True Positive	48
True Negative	1
False Positive	1
False Negative	0
Recall	100%

Table 10.8 reveals that a total of 50 photographs were checked for low brightness; 47 of those images were properly recognized, 2 of those images were mistakenly identified as forged, 2 of those images were correctly identified as non-forged, and 1 image was mistakenly missed but turned out to be forged.

The Recall Graph for Low Brightness images is shown in Figure 10.17:

The recall value of the suggested model for low brightness images is displayed in Figure 10.17 graph. The graph displays the greatest recall rate, which is recorded at 97.91%.

With the aid of evaluation metrics, Figure 10.18 compares the recall of the suggested model for images with bright colors and low brightness. The graph displays the greatest recall, which is 100% for bright color images and 97.11% for low brightness images.

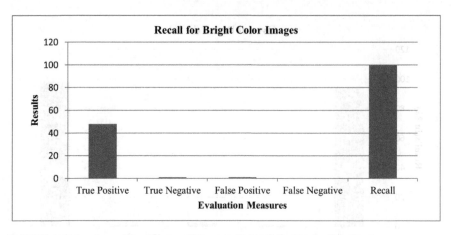

FIGURE 10.16 shows Recall value of Proposed Model for Bright Color Images.

TABLE 10.8
shows Recall for Low Brightness Images

PROPERTY	Low Brightness Images
True Positive	47
True Negative	0
False Positive	2
False Negative	1
Recall	97.91%

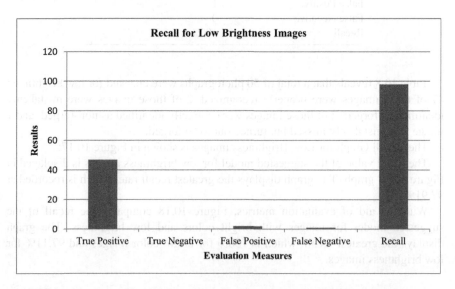

FIGURE 10.17 shows Recall value of Proposed Model for Low Brightness Images.

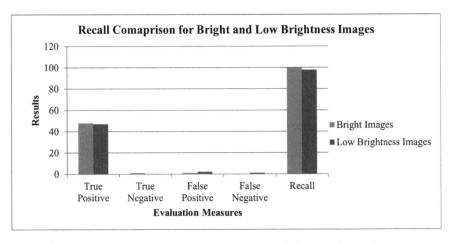

FIGURE 10.18 shows Recall comparison of Proposed Model for Bright Color and Low Brightness Images.

TABLE 10.9
shows evaluation measures for Prevalence

Condition of Image	Total images tested	True positive (TP)	True negative (TN)	False positive (FP)	False negative (FN)	Prevalence (in %)
Overall	100	95	1	3	1	96.00%

10.4.4.4 Prevalence

For convenience of interpretation, prevalence is expressed as percentages. It is computed using:

$$Prevalence = TP+FN/TP+FN+FP+TN$$

Table 10.9 demonstrates that a total of 100 photos were evaluated, of which 95 were found to be true positives—that is, correctly detected. There are a total of five failures: one true negative, one false negative, and three false positives. The graph in Figure 10.19 displays the overall prevalence:

The suggested model's overall prevalence is displayed in this graph. The maximum prevalence, represented in the graph at 96%, is documented.

There are a total of 100 photographs evaluated in Table 10.10, 50 of which are color images and the remaining 50 are low brightness images. 48 photos of bright color are successfully identified, 1 image is accurately identified as not-forged, and 1 image is mistakenly identified as forged. Figure 10.20 is the Bright Color image prevalence graph.

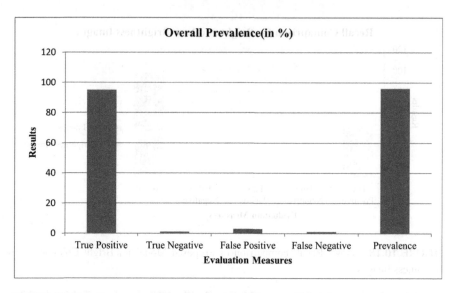

FIGURE 10.19 shows overall Prevalence value of Proposed Model.

TABLE 10.10
shows parameter values for Bright Color Images

PROPERTY	Bright Color Images
True Positive	48
True Negative	1
False Positive	1
False Negative	0
Prevalence	96%

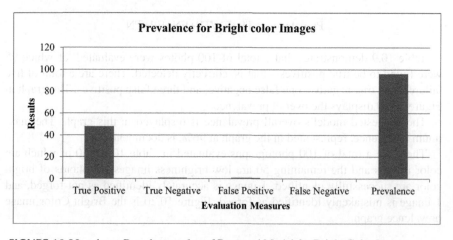

FIGURE 10.20 shows Prevalence value of Proposed Model for Bright Color Images.

Optimized Image Recognition for Smart Information Systems

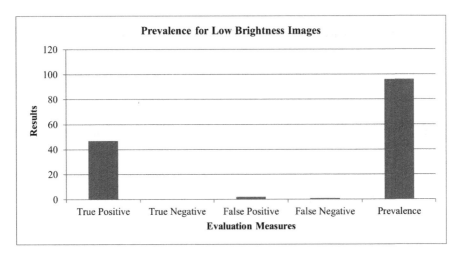

FIGURE 10.21   shows Prevalence value of Proposed Model for Low Brightness Images.

TABLE 10.11
shows recall for Low Brightness Images

PROPERTY	Low Brightness Images
True Positive	47
True Negative	0
False Positive	2
False Negative	1
Recall	96%

The suggested model's prevalence value for bright color images is displayed in Figure 10.20. The graph displays the highest documented prevalence, which is 96%.

Table 10.11 reveals that a total of 50 photographs were checked for low brightness; 47 of those images were properly recognized, 2 of those images were mistakenly identified as forged, 2 of those images were correctly identified as non-forged, and 1 image was mistakenly missed but turned out to be forged. Figure 10.21 is the Low Brightness image prevalence graph:

The suggested model's prevalence value for low brightness images is displayed in Figure 10.21. The maximum prevalence, represented in the graph at 96%, is documented.

With the use of evaluation metrics, the proposed model's prevalence comparison for bright color and low brightness images is displayed in Figure 10.22. The graph displays the greatest prevalence, which is 96% for bright color images and 96% for low brightness images.

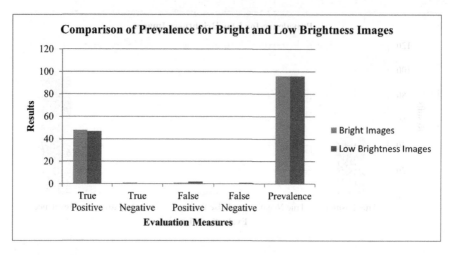

FIGURE 10.22 shows Prevalence comparison of Proposed Model for Bright Color and Low Brightness Images.

TABLE 10.12
compares the suggested model with the current model for a range of parameters

Parameters	SURF	SURF+SIFT
Sensitivity (Recall)	76%	98%
Precision (Positive Predictive Value)	77.00%	97.95%
Result Prevalence	76.00%	96.00%

10.5 COMPARISON OF PARAMETERS OF PROPOSED MODEL (SURF+SIFT) WITH EXISTING MODEL (SURF)

Based on the parameters listed in Table 10.12 below, this table compares the values of the current and proposed models:

The suggested approach, which combines SIFT and SURF, is contrasted in this table with SURF alone. The parameter graph is displayed in Figure 10.23 based on these parameters:

Figure 10.23 compares the suggested model with the current models for prevalence, sensitivity, and recall. The graph above displays the maximum values of Recall, Precision, and Prevalence, which are 98%, 97.95%, and 96%, respectively.

In Table 10.13, the proposed method which is combination of SIFT and SURF is compared with SURF alone for accuracy, the graph for accuracy is shown in Figure 10.24.

Figure 10.24 compares the suggested model with the current one for accuracy. The graph illustrates the highest accuracy, which is 96% recorded.

Optimized Image Recognition for Smart Information Systems 213

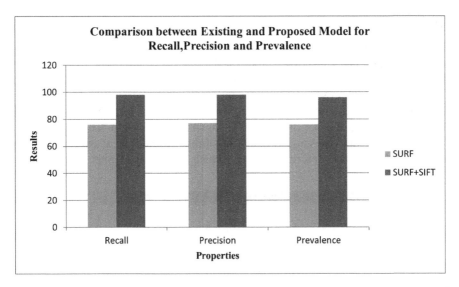

FIGURE 10.23 compares the proposed model with the current model for a range of parameters.

TABLE 10.13 shows comparison of proposed model with existing model for accuracy

Parameter	SURF	SIFT+SURF
Accuracy	76.00%	96.00%

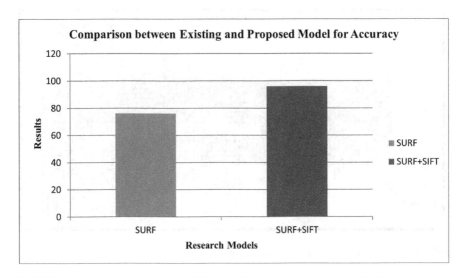

FIGURE 10.24 shows comparison of Proposed Model with existing model for accuracy.

10.6 APPENDIX I

The primary screen for the suggested model is depicted in the aforementioned image (Figure 10.25). There are two options on the main screen: choose image and run. First, upload the image by selecting the Select Image button. The results will be shown on a number of screens once you hit the RUN button.

Figure 10.26 illustrates how the SURF method operates. Here, SURF extracts the image's key points and descriptor vectors, which are displayed in the screenshot.

Figure 10.27 illustrates how the SIFT algorithm operates. Here, SIFT extracts the image's key points and descriptor vectors, which are displayed in the snapshot.

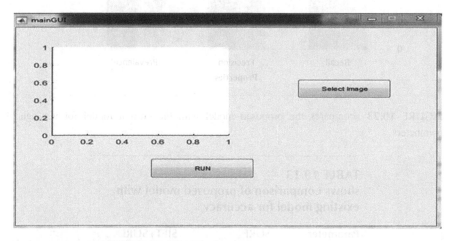

FIGURE 10.25 The simulation Scenario.

FIGURE 10.26 shows results of SURF Algorithm.

Optimized Image Recognition for Smart Information Systems 215

FIGURE 10.27   shows results of SIFT Algorithm.

FIGURE 10.28   shows results of Proposed (SURF+SIFT) Method.

The operation of the SURF and SIFT methods is depicted in Figure 10.28. Here, the image's descriptor vectors and key-point features are extracted using both SIFT and SURF. The above-listed main points from both SIFT and SURF are merged in this snapshot.

Figure 10.29 compares the accuracy of SURF, SIFT, and the combination of SURF and SIFT. This graph makes it evident that the suggested model is more accurate than SIFT and SURF by itself.

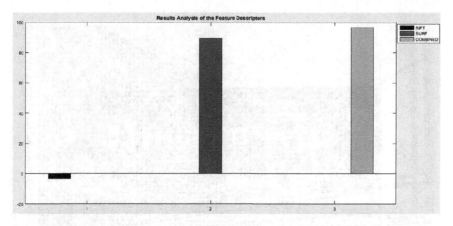

FIGURE 10.29 shows results analysis of Proposed (SURF+SIFT) Method.

REFERENCES

1. Alam, Sanawer, and Deepti Ojha. "A Literature study on Image forgery." *International Journal of Advance Research in Computer Science and Management Studies* 2, no. 10 (2014): 156–167.
2. Amerini, Irene, Lamberto Ballan, Roberto Caldelli, Alberto Del Bimbo, and Giuseppe Serra. "A sift-based forensic method for copy–move attack detection and transformation recovery." *IEEE Transactions on Information Forensics and Security* 6, no. 3 (2011): 1099–1110.
3. Amerini, Irene, Lamberto Ballan, Roberto Caldelli, Alberto Del Bimbo, Luca Del Tongo, and Giuseppe Serra. "Copy-move forgery detection and localization by means of robust clustering with J-Linkage." *Signal Processing: Image Communication* 28, no. 6 (2013): 659–669.
4. Ardizzone, Edoardo, Alessandro Bruno, and Giuseppe Mazzola. "Detecting multiple copies in tampered images." In 2010 IEEE International conference on image processing, pp. 2117–2120. IEEE, 2010.
5. Bayram, Sevinc, Husrev Taha Sencar, and Nasir Memon. "An efficient and robust method for detecting copy-move forgery." In 2009 IEEE International Conference on Acoustics, Speech and Signal Processing, pp. 1053–1056. IEEE, 2009.
6. Bianchi, Tiziano, and Alessandro Piva. "Detection of non-aligned double JPEG compression with estimation of primary compression parameters." In 2011 18th IEEE International Conference on Image Processing, pp. 1929–1932. IEEE, 2011.
7. Birajdar, Gajanan K., and Vijay H. Mankar. "Digital image forgery detection using passive techniques: a survey." *Digital Investigation* 10, no. 3 (2013): 226–245.
8. Jaberi, Maryam, George Bebis, Muhammad Hussain, and Ghulam Muhammad. "Accurate and robust localization of duplicated region in copy–move image forgery." *Machine Vision and Applications* 25 (2014): 451–475.
9. Cao, Gang, Yao Zhao, Rongrong Ni, and Alex C. Kot. "Unsharp masking sharpening detection via overshoot artifacts analysis." *IEEE Signal Processing Letters* 18, no. 10 (2011): 603–606.
10. Cao, Gang, Yao Zhao, Rongrong Ni, and Xuelong Li. "Contrast enhancement-based forensics in digital images." *IEEE Transactions on Information Forensics and Security* 9, no. 3 (2014): 515–525.

11. Christlein, Vincent, Christian Riess, Johannes Jordan, Corinna Riess, and Elli Angelopoulou. "An evaluation of popular copy-move forgery detection approaches." *IEEE Transactions on Information Forensics and Security* 7, no. 6 (2012): 1841–1854.
12. Cozzolino, Davide, Giovanni Poggi, and Luisa Verdoliva. "Efficient dense-field copy–move forgery detection." *IEEE Transactions on Information Forensics and Security* 10, no. 11 (2015): 2284–2297.
13. Farid, Hany. (2009). Exposing digital forgeries from JPEG ghosts. *IEEE Transactions on Information Forensics and Security*, 4(1), 154–160.
14. Farid, Hany. "Image forgery detection." *IEEE Signal Processing Magazine* 26, no. 2 (2009): 16–25.
15. Ferrara, Pasquale, Tiziano Bianchi, Alessia De Rosa, and Alessandro Piva. "Reverse engineering of double compressed images in the presence of contrast enhancement." In 2013 IEEE 15th International Workshop on Multimedia Signal Processing (MMSP), pp. 141–146. IEEE, 2013.
16. Gharibi, Fereshteh, Javad RavanJamjah, Fardin Akhlaghian, Bahram Zahir Azami, and Javad Alirezaie. "Robust detection of copy-move forgery using texture features." In 2011 19th Iranian Conference on Electrical Engineering, pp. 1–4. IEEE, 2011.
17. Zhou, Guojuan, and Dianji Lv. "An overview of digital watermarking in image forensics." In 2011 Fourth International Joint Conference on Computational Sciences and Optimization, pp. 332–335. IEEE, 2011.
18. Hashmi, Mohammad Farukh, Aaditya R. Hambarde, and Avinash G. Keskar. "Copy move forgery detection using DWT and SIFT features." In 2013 13th International Conference on *I*ntelligent *S*ystems *D*esign and *A*pplications, pp. 188–193. IEEE, 2013.
19. Hashmi, Mohammad Farukh, Vijay Anand, and Avinash G. Keskar. "A copy-move image forgery detection based on speeded up robust feature transform and Wavelet Transforms." In 2014 *I*nternational *C*onference on *C*omputer and *C*ommunication *T*echnology (ICCCT), pp. 147–152. IEEE, 2014.
20. Hsu, Chen-Ming, Jen-Chun Lee, and Wei-Kuei Chen. "An efficient detection algorithm for copy-move forgery." In 2015 10th Asia Joint Conference on Information Security, pp. 33–36. IEEE, 2015.
21. Jaberi, Maryam, George Bebis, Muhammad Hussain, and Ghulam Muhammad. "Accurate and robust localization of duplicated region in copy–move image forgery." *Machine Vision and Applications* 25 (2014): 451–475.
22. Jing, Li, and Chao Shao. "Image copy-move forgery detecting based on local invariant feature." *Journal of Multimedia* 7, no. 1 (2012): 167–175.
23. Khan, Saiqa, and Arun Kulkarni. "Robust method for detection of copy-move forgery in digital images." In 2010 International Conference on Signal and Image Processing, pp. 69–73. IEEE, 2010.
24. Li, Jian, Xiaolong Li, Bin Yang, and Xingming Sun. "Segmentation-based image copy-move forgery detection scheme." *IEEE Transactions on Information Forensics and Security* 10, no. 3 (2014): 507–518.
25. Li, Kunlun, Hexin Li, Bo Yang, Qi Meng, and Shangzong Luo. "Detection of image forgery based on improved PCA-SIFT." In Computer Engineering and Networking: Proceedings of the 2013 International Conference on Computer Engineering and Network (CENet2013), pp. 679–686. Springer International Publishing, 2014.
26. Li, Leida, Shushang Li, Hancheng Zhu, Shu-Chuan Chu, John F. Roddick, and Jeng-Shyang Pan. "An Efficient Scheme for Detecting Copy-move Forged Images

by Local Binary Patterns." *Journal of Information Hiding and Multimedia Signal Processing* 4, no. 1 (2013): 46–56.
27. Li, Weihai, and Nenghai Yu. "Rotation robust detection of copy-move forgery." In 2010 IEEE International Conference on Image Processing, pp. 2113–2116. IEEE, 2010.
28. Mahalakshmi, S. Devi, K. Vijayalakshmi, and S. Priyadharsini. "Digital image forgery detection and estimation by exploring basic image manipulations." *Digital Investigation* 8, no. 3–4 (2012): 215–225.
29. Muhammad, Najah, Muhammad Hussain, Ghulam Muhammad, and George Bebis. "Copy-move forgery detection using dyadic wavelet transform." In 2011 Eighth International Conference Computer Graphics, Imaging and Visualization, pp. 103–108. IEEE, 2011.
30. Panchal, P. M., S. R. Panchal, and S. K. Shah. "A comparison of SIFT and SURF." *International Journal of Innovative Research in Computer and Communication Engineering* 1, no. 2 (2013): 323–327.
31. Pan, Xunyu, and Siwei Lyu. "Region duplication detection using image feature matching." *IEEE Transactions on Information Forensics and Security* 5, no. 4 (2010): 857–867.
32. Qian, Ruohan, Weihai Li, Nenghai Yu, and Zhuo Hao. "Image forensics with rotation-tolerant resampling detection." In 2012 IEEE International Conference on Multimedia and Expo Workshops, pp. 61–66. IEEE, 2012.
33. Redi, Judith A., Wiem Taktak, and Jean-Luc Dugelay. "Digital image forensics: a booklet for beginners." *Multimedia Tools and Applications* 51 (2011): 133–162.
34. Sridevi, M., C. Mala, and Siddhant Sanyam. "Comparative study of image forgery and copy-move techniques." In Advances in Computer Science, Engineering & Applications: Proceedings of the Second International Conference on Computer Science, Engineering and Applications (ICCSEA 2012), May 25-27, 2012, New Delhi, India, Volume 1, pp. 715–723. Springer Berlin Heidelberg, 2012.
35. Sunil, Kumar, Desai Jagan, and Mukherjee Shaktidev. "DCT-PCA based method for copy-move forgery detection." In ICT and Critical Infrastructure: Proceedings of the 48th Annual Convention of Computer Society of India-Vol II: Hosted by CSI Vishakapatnam Chapter, pp. 577–583. Springer International Publishing, 2014.
36. Garg, Ankit, Aleem Ali, and Puneet Kumar. "Original Research Article A shadow preservation framework for effective content-aware image retargeting process." *Journal of Autonomous Intelligence* 6, no. 3 (2023): 245–252.
37. Wang, Junbin, Zhenghong Yang, and Shaozhang Niu. "Copy-move forgeries detection based on SIFT algorithm." *Proc. International Journal of Computer Science* 2, (2015): 567–570.
38. Yu, Liyang, Qi Han, and Xiamu Niu. "Feature point-based copy-move forgery detection: covering the non-textured areas." *Multimedia Tools and Applications* 75 (2016): 1159–1176.
39. Kumar, Sunil, and P. K. Das. "Copy-move forgery detection in digital images: progress and challenges." *International Journal on Computer Science and Engineering* 3, no. 2 (2011): 652–663.
40. Hamid, Irfan, Rameez Raja, Monika Anand, Vijay Karnatak, and Aleem Ali. "Comprehensive robustness evaluation of an automatic writer identification system using convolutional neural networks." (2023).
41. Sachdeva, Shaweta, and Aleem Ali. "Machine learning with digital forensics for attack classification in cloud network environment." *International Journal of System Assurance Engineering and Management* 13, no. Suppl 1 (2022): 156–165.

42. Ansari, Farah Jamal, and Aleem Ali. "A Comparison of the DCT JPEG and Wavelet Image Compression Encoder for Medical Images." In Contemporary Computing: 5th International Conference, IC3 2012, Noida, India, August 6-8, 2012. Proceedings 5, pp. 490–491. Springer Berlin Heidelberg, 2012.
43. Sachdeva, Shaweta, and Aleem Ali. "A hybrid approach using digital Forensics for attack detection in a cloud network environment." *International Journal of Future Generation Communication and Networking* 14, no. 1 (2021): 1536–1546.

11 Machine Learning-based Security Algorithms for Cloud Computing
A Comprehensive Survey

Shaweta Sachdeva, Aleem Ali, and
Ahmed A. Elngar

11.1 INTRODUCTION

The cloud environment's security has become increasingly crucial. Despite their robust security measures, well-known cloud service providers like Google and Amazon are not impervious to the many recorded cloud hacks. Cloud security can be broadly classified into five categories: information security, identity security, network security, infrastructure security, and software security. To assist defense mechanisms against many kinds of cloud threats, Cloud computing uses a service architecture called Machine Learning as a Service (MLaaS) Numerous Intrusion Detection Systems (IDS) have been built using machine learning methods, which have improved attack detection accuracy and guaranteed feature-free operation. Machine learning methods have affected daily life and altered practices. In recent years, they have also made notable progress in a number of areas.

The relevance of cloud environment security has increased significantly. Despite their robust security measures, well-known cloud service providers like Google and Amazon are not impervious to the many recorded cloud hacks. Cloud security can be classified into five categories: information security (IS), identity security, network security (NS), infrastructure security, and software security (SS). Cloud computing uses a service architecture called Machine Learning as a Service (MLaaS) to assist defense mechanisms against many kinds of cloud threats. Numerous Intrusion Detection Systems (IDS) have been built using machine learning techniques, which have improved the accuracy of detecting attacks and guaranteed the seamless continuation of operations.

11.2 THEORETICAL CONTEXT

This section provides a basic overview of machine learning and cloud computing by describing specifics.

Survey of Machine Learning Security Algorithms in Cloud Computing

11.2.1 Cloud Computing

A computer architecture known as "cloud computing" takes advantage of the internet to offer end users scalable, secure, and quantifiable services on demand. The numerous advantages of this paradigm account for its extensive variety of applications. In the current market, numerous cloud service providers—Amazon Web Services (AWS), Microsoft Azure, IBM Cloud, Google Cloud, Oracle Cloud, and Alibaba Cloud, to mention a few—offer a wide range of cloud services to their clients.

11.2.1.1 Characteristics of Cloud Computing

Cloud computing is characterized by five key elements. [1], [2] that are listed as follows:

I. **Self-service on demand:** Several cloud services are self-service on demand, meaning that customers can use them without the cloud service provider putting up any obstacles.
II. **Rapid elasticity:** Cloud-based apps with rapid flexibility can effectively manage changes in service demand, averting resource constraints and business interruptions.
III. **Measured service:** It allows customers to pay only for the services they use and to stop using the cloud at any moment. Users are billed for the cloud services according to their usage.
IV. **Broad network access:** Wide-ranging network connectivity guarantees that cloud services can be accessed from a range of thin clients, such as PDAs, laptops, desktops, and mobile devices.
V. **Resource pooling:** In order to meet the demands of several clients, cloud service providers always aggregate cloud computing resources such as memory storage, processors, and network bandwidth through a process known as resource pooling as shown in Figure 11.1.

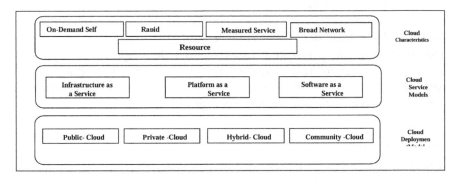

FIGURE 11.1 Framework of Cloud Computing.

11.2.1.2 Service Models
a. **Infrastructure as a Service (IaaS):** This approach provides the network, storage, processors, and virtual machines that are required to operate cloud applications. Because the service provider bears the entire cost of setup and maintenance, Infrastructure as a Service (IaaS) reduces those high prices.
b. **Platform as a Service (PaaS):** This approach gives developers a way to use the cloud infrastructure as a foundation for their app development. PaaS offers the technology, programming languages, and other tools needed for application development. Even if they have no authority over the cloud infrastructure supporting the program, end users nevertheless have complete control over it.
c. **Software as a Service (SaaS):** In this model, end users can access cloud-deployed applications over the internet through a number of clients. Users have no control over the cloud infrastructure or the application itself; they merely use the software [4].
d. **Hybrid Cloud:** A cloud architecture that blends multiple deployment techniques is called a hybrid cloud. Examples include VMware Cloud and similar installations.

11.2.1.3 Various Deployment Model
a. **Public Cloud Model:** Applications that have been made available via the public cloud can be used by any end user. Several well-known examples of public cloud services are Amazon EC2, Google App Engine, and Microsoft Azure.
b. **Private Cloud Model:** A private cloud guarantees a high degree of privacy and security without the involvement of other parties because it is exclusively used by particular private companies. A few examples of private clouds are Eucalyptus, Ubuntu Enterprise Cloud, Amazon Virtual Private Cloud, and Microsoft ECI data center.
c. **Community Cloud Model:** This deployment technique is applied when several organizations share cloud infrastructure [5]. Community cloud models can be administered in-house or through outsourcing, depending on the requirements. There are two examples: Microsoft Community Cloud and Google Apps for Government.

11.2.1.4 Classification of Cloud Attacks
The cloud computing paradigm is vulnerable to various types of assaults. These assaults may happen in various cloud service models, such as IaaS, PaaS, or SaaS, depending on their nature [6]. A few of the most well-known cloud attacks at the corresponding service models are shown in Figure 11.2.

11.2.2 MACHINE LEARNING
Arthur Samuel used the term "machine learning" to describe the process of learning from past experiences rather than by applying traditional programming techniques. In machine learning, a variety of methods are employed to create models that, after undergoing extensive training with large datasets including historical data from the past,

Survey of Machine Learning Security Algorithms in Cloud Computing

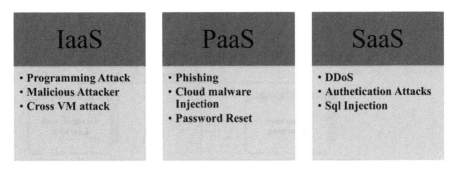

FIGURE 11.2 Classifications for Cloud Attacks.

can predict future outcomes. These algorithms form the basis of machine learning, and the type of problem to be solved determines which specific method to utilize. It takes a systematic approach to apply machine learning to problem solving. It starts with data collection and proceeds through data preparation, analysis, training, testing, and model deployment before the model is used in real-world situations [7], [8].

11.2.2.1 Types of Machine Learning

Machine learning (ML) can be broadly classified into several types based on the nature of the learning signal or feedback available to a learning system. Here are the primary types of machine learning:

Supervised Machine Learning: By mapping data to corresponding output goal values, supervised machine learning algorithms, using labeled datasets as training material, predict future outcomes. These algorithms take in incoming data and assign a suitable class based on large-scale, well-labeled datasets that have different class definitions from previous training. Supervised machine learning addresses two problem categories: classification for category target variables and regression for non-categorical, continuous target variables [9], [10].

Unsupervised Machine Learning: Algorithms for unsupervised machine learning are learned on datasets devoid of categories and labels. By sifting through large datasets, these algorithms automatically find and uncover data insights such as classes, categories, and trends. Association and clustering are the two main categories into which unsupervised machine learning falls.

Semi-Supervised Machine Learning: Semi-supervised machine learning algorithms address the limitations of both supervised and unsupervised approaches. These algorithms leverage both labeled and unlabeled datasets to train the machine learning model [11], [12].

Reinforcement Machine Learning: This type of machine learning relies on a feedback-based learning paradigm. Rather than being trained on supervised datasets, the agent is rewarded or penalized for making the right decisions based on its own experience. Figure 11.3 shows how machine learning algorithms are categorized.

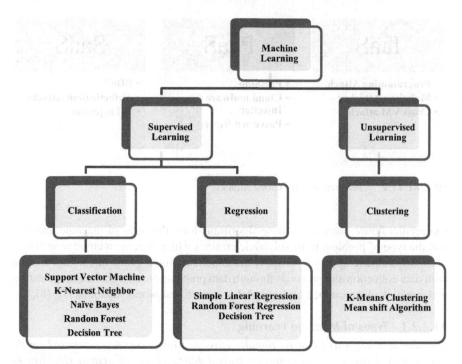

FIGURE 11.3 Various Classifications of Machine Learning Algorithms.

11.3 RESEARCH CONTEXT AND FOUNDATIONS

The phrase "machine learning" was first used by Arthur Samuel to refer to the process of learning from prior experiences as opposed to using conventional programming techniques. Many techniques are used in machine learning to build models that can forecast future results after undergoing substantial training with sizable datasets that contain historical data from the past. These algorithms are the cornerstone of machine learning, and the choice of which particular algorithm to use depends on the nature of the problem being solved. Applying machine learning to issue solving is a methodical process that begins with data collection and moves through steps including data preparation, analysis, training, testing, and model deployment before the model is put to use in real-world scenarios [7], [8].

Bagga et al. [14] provide an alternative security framework that combines the SVM machine learning technique with Network Function Virtualization (NFV) and Software Defined Network (SDN). This new approach becomes important since it provides defense against many attacks that target both SDN and NFV. The framework is constructed on two levels. The first tier, also referred to as the "security enforcement plane," is responsible for protecting the Internet of Things (IoT) against attacks that come from within and beyond. This level consists of three components: the Monitoring Agent (MA), the Control and Management Block (CMB), and the Infrastructure Block (IB). Runtime security policy setup falls within the purview of the "security orchestration plane," or second level. When compared to current methods, the proposed framework performs better in terms of detection rate, accuracy, False Rejection Rate (FRR), detection rate, and training duration.

Dey et al. [15] highlight the significance of data security in mobile cloud computing, particularly when taking heterogeneous networks into account. They recommend an intrusion detection system (IDS) that is capable of managing the intricate security specifications associated with these kinds of configurations. This intrusion detection system (IDS), which is based on the K-Means and DBSCAN machine learning algorithms, provides defense against a range of threats, such as distributed denial of service (DDoS) and man-in-the-middle (MITM). The technique employs a cluster-based training mechanism and classifies traffic using distance computations. The reduced complexity brought about by the infrequent rule updates is attributed to the improved correctness of the proposed IDS.

Alli et al. [16] present a machine learning-based safe offloading concept in a fog, cloud, and IoT multi-environment situation. For the purpose of storing Internet of Things data, the optimal fog node selection is determined using Particle Swarm Optimization (PSO). Next, data transfer to the cloud based on reinforcement learning is done. Only sensitive data is stored in a private cloud; non-sensitive data is not uploaded or stored there.

Rabbani et al. [17] use a machine learning-based strategy to monitor user behavior in the cloud is presented for the Cloud Service Provider (CSP). The technique employs a hybrid PSO-PNN (Particle Swarm Optimization and Probabilistic Neural Network) to discover unauthenticated users in the cloud. The results demonstrate the effectiveness of this hybrid technique, achieving excellent accuracy in terms of true positive rate, false negative rate, f-measure, and precision.

Hesamifard et al. [18] employ machine learning methodologies to enhance privacy safeguarding. The data encrypted with homomorphic encryption is used to train the neural network. The traditional sigmoid and ReLU (Rectified Linear Unit) activation functions of the neural network are replaced by exact polynomial approximations in this method. The proposed method demonstrates better privacy preservation than secure multi-party computation (SMC) and homomorphic encryption (HE).

Singh et al. [19] are able to achieve safe machine learning-based cloud data sharing by means of a mutual authentication method. This protocol offers strong defense against replay, DDoS, Man-in-the-Middle (MITM), and other cloud attacks. Data is encrypted by combining Schnorr's signature with Elliptic Curve Cryptography (ECC) and small-sized keys. For threat categorization or attack identification, a voting classifier is employed, and the Pro Verif tool results demonstrate the exceptional accuracy of the suggested methodology.

Salman et al.'s research paper [20] suggested employing machine learning in conjunction with an intrusion detection system to thwart various kinds of cloud attacks in a multi-cloud environment. The proposed intrusion detection system makes use of linear regression and supervised machine learning algorithms in random forests. This approach has the extra advantage of classifying threats in addition to identifying cloud threats because it follows a distinct step-by-step process. 99.0% classification accuracy and 93.6% threat detection accuracy were attained.

Chiba et al. [21] offer an intrusion detection system based on a deep neural network that integrates genetic and simulated annealing processes. The enhanced genetic algorithm used in this method minimizes execution time and convergence. The search phase of the genetic algorithm is simultaneously optimized using the simulated annealing technique. These methods improve the system's overall performance by strengthening the deep neural network's activation function and feature selection, among other aspects.

Khilar et al. [22] offer a machine learning-based authentication system that aims to grant access to cloud services exclusively to confirmed users. This proposed solution prevents unauthorized access to cloud resources by streamlining the authorization procedure for cloud users. It thus boosts service providers' confidence in end users, which contributes to an improvement in data security as a whole. Comparative results demonstrate that the proposed solution outperforms traditional mechanisms for user access to cloud resources in terms of Mean Absolute Error (MAE), time, recall, precision, and f1-score.

Aljamal et al. [23] have suggested an enhanced accuracy Intrusion Detection System (IDS) that is based on machine learning. In the hybrid model, machine learning techniques such as K-Means clustering and SVM for classification are incorporated at the cloud hypervisor level. This hybrid approach finds anomalies in the network by analyzing network traffic, removing unwanted features from the dataset, clustering the data using K-Means, and classifying requests as malicious or legal using SVM.

Sethi et al. [24] enhance cloud security by utilizing an intrusion detection system (IDS) that is based on reinforcement learning. This method addresses a significant problem with low False Positive Rate (FPR), or imprecise categorization accuracy, that traditional IDS for cloud security encounters. The proposed model comprises three primary components: the agent network, which distinguishes between legitimate and malicious requests, the host network, which guards against attacks on virtual machines, and the administration network, which enables administrators to stop affected virtual machines.

Chkirbenet et al. [25] introduced the "EIDS" traffic analysis technique, which combines machine learning-based intrusion detection with other techniques, with the goal of enhancing cloud security. This technique raises the performance of the intrusion detection system by categorizing attacks according to prior and current decision-making comparisons. The detection rate of the supervised learning classifier rises by 24%, or nearly 90%, suggesting that the security of the IDS is reinforced.

In a different study, Chkirbene et al. [26] proposed two models of trust-based intrusion detection systems (IDS): the Trust-based Intrusion Detection Classification Model (TIDCS) and its accelerated form (TIDCS-A). By focusing on dimensionality reduction, the former allows the UNSW dataset machine learning approach to process just the most significant data. The task of detecting anomalies falls to the later model. The simulation's results demonstrate the accuracy with which the two proposed models—which employ the machine learning algorithms TIDCS and TIDCS-A—can classify and identify attacks.

The topic of data manipulation comes up more and more when talking about distributed cloud systems and machine learning. To solve this problem and guarantee data integrity in distributed cloud systems, Zhao et al. [27] developed the DML-DIV (Distributed Machine Learning Data Integrity Verification) verification methodology. According to simulation results, the proposed DML-DIV strategy performs better than the existing comparative methods, particularly in terms of privacy protection and defense against forgery and tampering attacks as shown in Table 11.1.

TABLE 11.1
Summary of related work

Sr.No	ML Algorithm Used	Proposed Approach	Dataset
1	Decision Tree Algo. [13]	Weight optimization is the basis for the use of IDS (Intrusion Detecting System).	UNSW
2	Support Vector Machine Algo [14]	AI framework built by combining several ML, NFV, and SDN methodologies.	NSL-KDD
3	DBSCAN [15] and K-Means Algo	Different traffic filtrations are carried out using clustering as a training method and distance calculation.	Multiple datasets
4	Reinforcement Learning algo [16]	In a fog-cloud-IoT environment, the Neuro-Fuzzy System (NFS) selects the secure fog node via PSO-assisted secure data offloading.	Multiple datasets
5	Multilayer Neural Network (MNN) algo [17]	There are several methods for locating undesired users in the cloud using PSO and PNN.	UNSW-NB15
6	Deep Neural Network (DNN) algo [18]	Several methods for training neural networks using encrypted data and precise NN activation functions.	Crab, Fertility and Climate Dataset
7	LR (Linear Regression), KNN (K-Nearest neighbor) algorithms [19]	ECC and voting classifier combined for mutual authentication in a multi-cloud setting.	CICD
8	Linear Regression (LR), Random Forest (RF) [20]	An intrusion detection system using machine learning to identify threats in a multi-cloud context.	UNSW
9	K-Means clustering, SVM classification [23]	The hybrid model is in charge of investigating network traffic, removing undesired features from datasets, using the K-Means algorithm to cluster data, and using SVM to distinguish between legitimate and fraudulent requests.	UNSW-NB15

(*continued*)

TABLE 11.1 (Continued)
Summary of related work

Sr.No	ML Algorithm Used	Proposed Approach	Dataset
10	Random Forest, Quadratic Discriminant Analysis, K-Nearest Neighbors, Gaussian Naïve Bayes (GNB) and AdaBoost [24]	Using a host, agent, and administrator network, the Deep Reinforcement Learning model predicts which virtual machines would be impacted and disables them.	UNSW-NB15
11	Linear Regression (LR) [25]	Used for traffic analysis, where machine learning is used to analyze judgments made in the past and present.	UNSW-NB-15
12	Decision Tree and Random Forest [26]	Classification TIDCS and detection TIDCS-A models for IDS are based on machine learning algorithms.	NSL-KDD, UNSW
13	K-Nearest Neighbor, Decision Tree, Logistic Regression, Naïve Bays [22]	By applying a machine learning approach, the user's authorization is increased to improve the security of the cloud resources.	User Dataset

11.4 FUTURE WORKS

Our analysis shows that there have been relatively few security lapses on data in transit when machine learning has been successfully applied to data. However, the application of machine learning in cyber security and network security domains such as intrusion detection remains challenging due to evolving security attacks on data in transit. Thus, more research into possible protection mechanisms is required to fully understand the risks associated with security lapses and data privacy in machine learning. Some of the most crucial concerns with protecting the security and privacy of data in machine learning need to be considered for the following scenarios.

Applying machine learning would undoubtedly lead to the emergence of new security risks. This survey study looks at algorithm strategies based on machine learning (ML) and offers practical validation.

- Data privacy is critical when employing machine learning for security, requiring continuous improvements in machine methodologies. It would be challenging to conduct more research on sophisticated, reasonably priced privacy technology.
- New machine learning applications are needed for cyber and network security.

To create highly standardized procedures, a clearly defined security evaluation must be carried out. Further investigation into each of the aforementioned scenarios is recommended in order to provide a high level of security and privacy of data in machine learning.

11.5 CONCLUSIONS

Since it is essential to guarantee the security of client data in the cloud, researchers have been looking at a number of technologies and security methods to improve the security of the cloud ecosystem [28], [31]. Machine learning has become an essential tool to provide accurate and automatic defense against known and unforeseen cloud threats. The primary objective of this book chapter is to present a current overview of research projects in the area of cloud security based on machine learning. Our long-term goal is to develop an intrusion detection system that can deliver cloud data with enhanced security accuracy by utilizing machine learning techniques that have been improved and optimized.

REFERENCES

[1] Alouffi, B., Hasnain, M., Alharbi, A., Alosaimi, W. "A systematic literature review on cloud computing security: Threats and mitigation strategies". *IEEE Access*, vol. 9, pp. 57792–57807, 2021.

[2] Abdulsalam, Hedabou M. "Security and privacy in cloud computing: Technical review". *Future Internet*, vol. 14, p. 11, 2022.

[3] George, S.S., Pramila, R.S. "A review of different techniques in cloud computing". *Materials Today Proceedings*, vol. 46, pp. 8002–8008, 2021.

[4] Attaran, M., Woods, J. "Cloud computing technology: Improving small business performance using the Internet". *Journal of Small Business & Entrepreneurship*, vol. 13, pp. 94–106, 2018.

[5] Basu, S., Bardhan, A., Gupta, K., Saha, P., Pal, M., Bose, M., Basu, K., Chaudhury, S., Sarkar, P. "Cloud computing security challenges & solutions-A survey". In: *Annual Computing and Communication Workshop and Conference(CCWC)*, 2018.

[6] Dwivedi, R.K., Saran, M., Kumar, R. "A Survey on Security over Sensor-Cloud". In: *2019 9th International Conference on Cloud Computing, Data Science & Engineering (Confluence), Noida, India* (pp. 31–37), 2019.

[7] Butt, U.A., Mehmood, M., Shah, S.B.H., Amin, R., Shaukat, M.W., Raza, S.M., Suh, D.Y., Piran, M.J. "A review of machine learning algorithms for cloud computing security". *Electronics*, vol. 9, p. 1379, 2020.

[8] Sarker, I.H. "Machine Learning: Algorithms, Real-World Applications and Research Directions". *SN Computer Science*, vol. 2, no. 3, p. 160, 2021.

[9] Alzubi, J., Nayyar, A., Kumar, A." Machine Learning from Theory to Algorithms: An Overview". Journal of Physics: Conference Series, Volume 1142, Second National Conference on Computational Intelligence 2018, Bangalore, India.

[10] Baraneetharan, E. "Role of machine learning algorithms intrusion detection in WSNs: A survey". *Journal of Information Technology and Digital World*, Vol. 2, pp. 161–173, 2020.

[11] Saranyaa, T., Sridevi, S., Deisy, C., Chung, T.D., Khan, M.K.A. "Performance analysis of machine learning algorithms in intrusion detection system: A review". *Procedia Computer Science*, vol. 171, pp. 1251–1260, 2020.

[12] Sen, P.C., Hajra, M., Ghosh, M. "Supervised classification algorithms in machine learning: A survey and review". *Emerging Technology in Modelling and Graphics. Advances in Intelligent Systems and Computing*, vol. 937, pp. 99–111, 2019.

[13] Chkirbene, Z., Erbad, A., Hamila, R., Gouissem, A., Mohamed, A., Hamdi, M. "Machine learning based cloud computing anomalies detection". *IEEE Network*, vol. 34, pp. 178–183, 2020.

[14] Bagaa, M., Taleb, T., Bernabe, J.B., Skarmeta, A. "A machine learning security framework for Iot systems". *IEEE Access*, vol. 8, pp. 114066–114077, 2020.

[15] Dey, S., Ye, Q., Sampalli, S. "A machine learning based intrusion detection scheme for data fusion in mobile clouds involving heterogeneous client networks". *Information Fusion*, vol. 49, pp. 205–215, 2019.

[16] Alli, A.A., Alam, M.M.: SecOFF-FCIoT. "Machine learning basedsecure offloading in Fog-Cloud of things for smart city applications". *Internet of Things*, vol. 7, 2019.

[17] Rabbani, M., Wang, Y.L., Khoshkangini, R., Jelodar, H., Zhao, R., Hu, P. "A hybrid machine learning approach for malicious behavior detection and recognition in cloud computing". *Journal of Network and Computer Applications*, Online Journal, vol. 151, 2020.

[18] Hesamifard, E., Takabi, H., Ghasemi, M., Jones, C. "Privacy-preserving Machine Learning in Cloud". *Cloud Computing Security Workshop*, pp. 39–43, 2017.

[19] Singh, A.K., Saxena, D. "A cryptography and machine learning based authentication for secure data-sharing in federated cloud services environment". *Journal of Applied Security Research*, 2021.

[20] Salman, T., Bhamare, D., Erbad, A., Jain, R., Samaka, M. "Machine learning for anomaly detection and categorization in multi-cloud environments". *IEEE 4th International Conference on Cyber Security and Cloud Computing*, 2017.

[21] Chiba, Z., Abghour, N., Moussaid, K., Elomri, A., Rida, M. "Intelligent approach to build a Deep Neural Network based IDS for cloud environment using combination of machine learning algorithms". *Computers & Security*, vol. 86, pp. 291–317, 2019.

[22] Khilar, P.M., Chaudhari, V., Swain, R.R. "Trust-based access control in cloud computing using machine learning". *Cloud Computing for Geo spatial Big Data Analytics*, pp. 55–79, 2018.

[23] Aljamal, I., Tekeoğlu, A., Bekiroglu, K., Sengupta, S. "Hybrid intrusion detection system using machine learning techniques in cloud computing environments". *IEEE 17th International Conference on Software Engineering Research, Management and Applications (SERA)*, 2019.

[24] Sethi, K., Kumar, R., Prajapati, N., Bera, P. "Deep reinforcement learning based intrusion detection system for cloud infrastructure". *International Conference on Communication Systems & Networks(COMSNETS)*, 2020.

[25] Chkirbene, Z., Erbad, A., Hamila, R. "A combined decision for secure cloud computing based on machine learning and past information". *IEEE Wireless Communications and Networking Conference (WCNC)*, 2019.

[26] Chkirbene, Z., Erbad, A., Hamila, R., Mohamed, A., Guizani, M., Hamdi, M.: TIDCS. "A dynamic intrusion detection and classification system based feature selection". *IEEE Access*, vol. 8, pp. 95864–95877, 2020.

[27] Zhao, X., Jiang, R. "Distributed machine learning oriented data integrity verification scheme in cloud computing environment". *IEEE Access*, vol. 8, pp. 26372–26384, 2020.

[28] Sachdeva, S., Ali, A. "A Hybrid approach using digital Forensics for attack detection in a cloud network environment". *International Journal of Future Generation Communication and Networking*, vol. 14, no. 1, pp. 1536–1546, 2021.

[29] Sachdeva, S., Ali, A. "Machine learning with digital forensics for attack classification in cloud network environment". *International Journal of System Assurance Engineering Management*, Springer, 2021. https://doi.org/10.1007/s13198-021-01323-4

[30] Sachdeva, S., Ali, A., Khan, S. "Secure and privacy issues in telemedicine: Issues, solutions, and standards pp. 321–331". Chapter · January 2022. https://doi.org/10.1007/978-3-030-99457-0_19

[31] Sachdeva, S., Ali, Aleem, Khalid, Salman. "Telemedicine in healthcare system: A discussion regarding several practices pp. 295–310". Chapter · January 2022 DOI: 10.1007/978-3-030-99457-0_19

12 Securing Information in Transit

Leveraging AI/ML for Robust Data Protection

Navjot Singh Talwandi and Kulvinder Singh

12.1 INTRODUCTION TO INFORMATION SECURITY DURING TRANSIT

Information security during transit refers to the protection of data and information while it is being transmitted or transferred from one location to another. In today's interconnected world, where data is constantly being exchanged between systems, networks, and devices, ensuring the security of information during transit is crucial to prevent unauthorized access, interception, or tampering [1]. During transit, data can be vulnerable to various threats, including interception by malicious actors, data breaches, unauthorized access, and data corruption. Therefore, implementing robust security measures is essential to safeguard the confidentiality, integrity, and availability of information.

Key aspects of information security during transit include [2]:

Encryption: Encryption is a fundamental security measure that converts data into an unreadable format during transmission. It ensures that even if the data is intercepted, it remains unintelligible to unauthorized individuals. Implementing strong encryption algorithms and using secure protocols, such as Transport Layer Security (TLS) or Secure Sockets Layer (SSL), helps protect data during transit.

Secure Protocols: Using secure protocols for data transmission is crucial to prevent unauthorized access or interception. Secure protocols establish a secure channel between the sender and receiver, ensuring the confidentiality and integrity of the data. Examples of secure protocols include HTTPS for web communication and Secure File Transfer Protocol (SFTP) for file transfers.

Authentication and Access Control: Implementing strong authentication mechanisms, such as usernames, passwords, or multi-factor authentication, helps ensure that only authorized individuals can access and transmit data. Access control measures, such as role-based access control (RBAC), should be in place to restrict access to sensitive information during transit.

Data Integrity: Data integrity ensures that data remains unchanged and uncorrupted during transit. Implementing mechanisms like checksums or digital signatures helps verify the integrity of data at the receiving end. Hash functions can be used to generate unique identifiers for data, allowing the recipient to verify its integrity.

Intrusion Detection and Prevention Systems (IDPS): IDPS can monitor network traffic during transit and detect any suspicious or malicious activities. These systems can help identify and prevent unauthorized access attempts, network attacks, or data breaches during transmission.

Virtual Private Networks (VPNs): VPNs create secure and encrypted connections over public networks, such as the internet. By establishing a private network tunnel, VPNs ensure the confidentiality and integrity of data transmitted between remote locations or across public networks.

Data Loss Prevention (DLP): DLP solutions can help prevent accidental or intentional data leaks during transit. These solutions monitor and control data transfers, ensuring that sensitive information is not transmitted outside authorized channels or to unauthorized recipients.

Regular Updates and Patching: Keeping software, operating systems, and network devices up to date with the latest security patches and updates is crucial to address vulnerabilities and protect against emerging threats during transit. By implementing these security measures, organizations can mitigate the risks associated with data transmission and ensure the secure exchange of information between systems, networks, and devices. Information security during transit is a critical component of overall data protection and should be a priority for organizations to maintain the confidentiality, integrity, and availability of their data.

12.2 THREAT LANDSCAPE IN DATA TRANSMISSION

The threat landscape in data transmission encompasses various risks and vulnerabilities that data faces during its movement across networks or channels (Figure 12.1). Data can be intercepted or eavesdropped upon while in transit, especially if transmitted over unsecured or public networks. Hackers can intercept sensitive information, leading to data breaches or unauthorized access. Attackers can position themselves between the sender and receiver, intercepting and potentially altering the data being transmitted. This enables them to steal data, inject malicious content, or manipulate information without detection.

During transmission, data integrity can be compromised through unauthorized alterations. Hackers may modify or tamper with data packets, leading to corrupted or falsified information upon reaching the intended recipient.

Malicious actors can flood networks or systems with excessive traffic, caus- ing disruptions and rendering services inaccessible. These attacks disrupt data transmission and affect the availability of services. Attackers employ tactics like phishing emails or spoofed websites to trick users into revealing sensitive information, such as login credentials or financial details, which can compromise data during transmission. Inadequate encryption protocols or weak encryption keys used during data transmission can make it susceptible to decryption by unauthorized parties, compromising

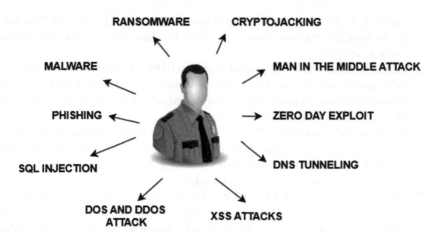

FIGURE 12.1 Threat landscape in data transmission.

confidentiality. Employees or individuals within an organization might misuse their access privileges to intercept or leak sensitive data during transmission, posing a significant internal threat. Outdated or insecure network protocols and technologies can create vulnerabilities that attackers exploit to gain unauthorized access to transmitted data.

Understanding and mitigating these threats involves employing robust encryption methods, implementing secure communication protocols, employing intrusion detection systems, conducting regular security audits, and fostering a culture of cybersecurity awareness among users.

12.3 FUNDAMENTALS OF AI/ML IN INFORMATION SECURITY

12.3.1 Overview of AI and ML Applications in Cybersecurity

Artificial Intelligence (AI) and Machine Learning (ML) play pivotal roles in revolutionizing cybersecurity by offering advanced capabilities to detect, prevent, and respond to cyber threats. AI/ML algorithms analyze vast amounts of data to identify patterns indicative of cyber threats. They can recognize anomalies in network traffic, behaviors, or system activities that signify potential attacks, aiding in early threat detection. ML models use historical data to predict potential vulnerabilities or risks within systems [3]. By analyzing patterns and trends, they help cybersecurity teams proactively address weaknesses before they're exploited. AI-driven systems establish baseline behaviors for networks, devices, or users. They then detect deviations from these norms, flagging unusual activities that might indicate a security breach or unauthorized access. ML algorithms analyze file structures, code behavior, and characteristics to identify new strains of malware or previously unseen threats. They can adapt and learn from new samples to improve malware detection capabilities. AI-powered systems enable automated responses to cyber threats. They can autonomously isolate infected systems, contain breaches, and initiate incident response measures, reducing response time and limiting damage. ML algorithms monitor and analyze user activities to detect suspicious behavior. By identifying deviations from normal user behavior, UBA helps prevent insider threats or account takeovers.

AI/ML techniques are employed to detect and thwart phishing attempts by analyzing email content, sender behavior, or website authenticity. They enhance fraud detection by identifying fraudulent transactions or activities. AI/ML technologies automate routine security tasks such as log analysis, patch management, and security configuration, reducing the burden on cybersecurity teams and enhancing overall efficiency. AI/ML continuously learn and adapt to evolving threats. They improve security measures by dynamically adjusting defense mechanisms based on real-time threat intelligence and changing attack patterns. AI assists in processing and analyzing vast amounts of threat intelligence data, providing cybersecurity professionals with actionable insights and recommendations for effective decision-making.

AI and ML applications in cybersecurity significantly bolster defense capabilities, enabling proactive threat mitigation, rapid incident response, and adaptive security measures in the ever-evolving landscape of cyber threats [4].

12.4 ENCRYPTION AND DECRYPTION TECHNIQUES

12.4.1 In-depth Exploration of Encryption Methodologies and Their Role in Safeguarding Data in Transit

Encryption methodologies play a pivotal role in securing data during transmission across networks or channels. Here's an in-depth exploration of various encryption methods and their significance in safeguarding data in transit:

Symmetric Encryption: This method uses a single key for both encryption and decryption. The sender encrypts the data using the key, and the recipient decrypts it with the same key. While efficient, securely sharing the key poses a challenge.

Asymmetric Encryption (Public-Key Encryption): Asymmetric encryption involves a pair of keys—a public key used for encryption and a private key for decryption. The sender uses the recipient's public key to encrypt the data, and only the recipient, holding the corresponding private key, can decrypt it.

Hash Functions: Hashing converts data into a fixed-size string of characters, known as a hash value or digest. It's a one-way process, making it infeasible to reverse-engineer the original data from the hash. Hash functions ensure data integrity by verifying if data has been tampered with during transmission.

Transport Layer Security (TLS) and Secure Sockets Layer (SSL): TLS and SSL protocols establish secure encrypted connections between clients and servers. They use a combination of asymmetric and symmetric encryption for secure data exchange, commonly used in web browsing and online transactions.

Virtual Private Networks (VPNs): VPNs use encryption to create a secure, encrypted tunnel for data transmission over public networks. They employ various encryption protocols to protect data from interception or eavesdropping.

Advanced Encryption Standard (AES): AES is a widely adopted symmetric encryption algorithm used for securing sensitive data. It's known for its robustness and efficiency, employing varying key lengths (128, 192, or 256 bits) for encryption.

Elliptic Curve Cryptography (ECC): ECC is an asymmetric encryption technique known for its ability to provide strong security with shorter key lengths compared to other methods. It's particularly useful in resource-constrained environments like IoT devices or mobile applications.

Quantum Encryption: Still in experimental stages, quantum encryption leverages principles of quantum mechanics for ultra-secure communication. Quantum key distribution (QKD) uses quantum properties to ensure the secrecy of encryption keys.

Encryption methodologies play a critical role in safeguarding data in transit by ensuring confidentiality, integrity, and authenticity. They protect sensitive information from unauthorized access or modifications, mitigate the risks of interception or tampering, and provide a secure foundation for data exchange in today's interconnected digital world. Implementing robust encryption protocols is fundamental in ensuring the privacy and security of transmitted data [5].

12.4.2 AI/ML-DRIVEN ADVANCEMENTS IN ENCRYPTION ALGORITHMS FOR ENHANCED SECURITY

AI and Machine Learning (ML) have been instrumental in advancing encryption algorithms, contributing to enhanced security in several ways (Figure 12.2):

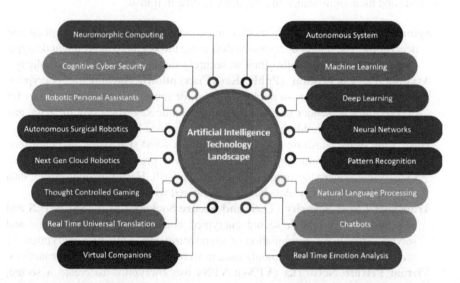

FIGURE 12.2 AI/ML-driven advancements in encryption algorithms for enhanced security.

Securing Data in Transit with AI/ML Solutions

Improved Key Management: AI/ML algorithms aid in creating and managing encryption keys more efficiently. They analyze patterns and usage behaviors to generate stronger, unique keys and manage their lifecycle, reducing vulnerabilities associated with weak or reused keys [6].

Enhanced Cryptanalysis and Vulnerability Detection: ML techniques assist in analyzing encryption algorithms to identify potential weaknesses or vulnerabilities. By simulating attacks and analyzing patterns in ciphertext, ML helps in fortifying algorithms against emerging threats.

Adaptive Encryption Techniques: AI-driven systems adapt encryption techniques based on real-time threat intelligence. They dynamically adjust encryption methods or key strengths in response to evolving threats, ensuring robust protection against sophisticated attacks.

Quantum-Safe Cryptography: With the advent of quantum computing, AI/ML plays a role in developing quantum-resistant encryption algorithms. ML-driven research aids in designing cryptographic systems that remain secure against quantum-based attacks, ensuring future-proof security.

Behavioral Encryption and Access Control: AI analyzes user behavior patterns to strengthen access control and encryption. ML algorithms detect anomalies in user behaviors and dynamically adjust encryption levels or access permissions based on risk assessments, preventing unauthorized access.

Predictive Encryption and Threat Mitigation: AI/ML models predict potential threats and encryption requirements based on historical data and real-time network analysis. This proactive approach helps in preemptively implementing stronger encryption measures to mitigate risks.

AI-Optimized Encryption Performance: ML algorithms optimize encryption performance by fine-tuning algorithms for speed and efficiency without compromising security. This optimization ensures faster encryption and decryption processes, enabling secure real-time data transmission.

Neural Cryptography: Exploring neural network architectures for cryptography is a novel area where AI and ML contribute. Neural networks are being studied for their potential to create encryption models that are resistant to adversarial attacks.

These AI/ML-driven advancements in encryption algorithms pave the way for more robust, adaptive, and resilient security measures. By leveraging machine learning capabilities, encryption methods can evolve to counteract emerging threats, adapt to changing environments, and provide enhanced protection for sensitive data during transmission and storage.

12.5 AI-POWERED THREAT DETECTION IN TRANSIT

AI-powered threat detection in transit refers to the utilization of Artificial Intelligence (AI) technologies to identify and mitigate potential security threats faced by data during its transmission across networks or channels. Here's an overview of how AI contributes to threat detection in transit:

Anomaly Detection: AI algorithms analyze network traffic patterns, system behaviors, and data transmissions to establish baselines of normal activities. Any deviations or anomalies from these patterns, such as unusual data access or transfer, can trigger alerts for potential threats like intrusions or data breaches.

Behavioral Analysis: AI-powered systems continuously learn and analyze user behaviors, identifying suspicious or unauthorized activities during data transmission. By recognizing abnormal behavior, such as unexpected access attempts or unusual data transfer volumes, AI helps flag potential threats.

Pattern Recognition and Prediction: AI/ML models employ pattern recognition to detect known attack signatures or indicators of compromise during data transmission. Additionally, predictive analytics based on historical data aid in anticipating and mitigating potential threats before they escalate.

Malware Detection: AI algorithms analyze file structures, code behavior, and characteristics to identify known and unknown malware during data transmission. ML-driven malware detection techniques can swiftly recognize and block malicious files or activities.

Threat Intelligence Integration: AI systems leverage threat intelligence feeds and databases to stay updated on emerging threats and attack patterns. By integrating this intelligence, AI-powered threat detection can proactively identify and respond to evolving threats during data transmission.

Real-time Monitoring and Response: AI continuously monitors network traffic and data transmissions in real time. It swiftly detects suspicious activities, triggers alerts, and initiates automated responses or security protocols to mitigate threats as they occur.

Adaptive Security Measures: AI adapts to changing threat landscapes and evolving attack techniques. It learns from new threats and adjusts threat detection algorithms, enabling adaptive security measures that stay ahead of emerging risks during data transmission.

Contextual Analysis: AI considers contextual information surrounding data transmissions, such as user roles, access privileges, and geographical locations. By analyzing context, AI-driven threat detection can differentiate between normal and suspicious activities more accurately.

AI-powered threat detection in transit enhances the ability to identify, analyze, and respond to potential security risks during data transmission. It offers proactive defense mechanisms that enable quicker threat identification, reducing the likelihood and impact of cyber threats on sensitive data in transit.

12.6 SECURE COMMUNICATION PROTOCOLS

12.6.1 OVERVIEW OF SECURE COMMUNICATION STANDARDS (E.G., SSL/TLS) AND THEIR INTEGRATION WITH AI/ML FOR IMPROVED DATA PROTECTION

Secure communication standards like SSL (Secure Sockets Layer) and its successor TLS (Transport Layer Security) are crucial protocols used to establish encrypted connections between clients and servers, ensuring data confidentiality and integrity

Securing Data in Transit with AI/ML Solutions

during transmission. Here's an overview of these standards and their integration with AI/ML for enhanced data protection:

SSL/TLS Encryption: SSL and TLS protocols encrypt data transmitted over networks, preventing unauthorized access or eavesdropping. They use asymmetric and symmetric encryption techniques to secure communication channels between devices or applications.

AI/ML Integration for Enhanced Encryption:

- Enhanced Key Management: AI/ML algorithms aid in optimizing key management for SSL/TLS encryption. They assist in generating, distributing, and managing encryption keys more securely, ensuring stronger protection against key-related vulnerabilities.
- Anomaly Detection and Threat Prevention: AI/ML-powered systems complement SSL/TLS by monitoring network traffic patterns and behaviors. They detect anomalies or suspicious activities that might bypass traditional encryption methods, enhancing threat prevention capabilities.
- Behavioral Analysis for Access Control: AI analyzes user behaviors and access patterns within SSL/TLS-secured communications. ML-driven behavioral analysis helps in identifying unusual access attempts or abnormal data transfer activities, enhancing access control measures.
- Predictive Encryption and Adaptive Security: AI/ML models predict potential threats or vulnerabilities within SSL/TLS-secured connections based on historical data. They facilitate adaptive encryption adjustments or security measures to proactively mitigate risks and maintain robust protection.

Continuous Improvement and Adaptation:

- Threat Intelligence Integration: AI systems integrated with SSL/TLS leverage threat intelligence to stay updated on emerging threats or vulnerabilities. This integration enhances SSL/TLS by providing real-time insights into evolving attack patterns, enabling adaptive security measures.
- Performance Optimization: AI/ML-driven optimization helps enhance the performance of SSL/TLS encryption. By analyzing and fine-tuning encryption processes, AI assists in improving encryption efficiency without compromising security.

Quantum-Safe Cryptography Exploration:

- Preparing for Future Threats: AI/ML contributes to research and development in quantum-safe cryptography for SSL/TLS. It explores new encryption methods resilient to potential threats posed by quantum computing, ensuring future-proof security for encrypted communications. The integration of AI/ML with secure communication standards like SSL/TLS fortifies data protection by enhancing encryption processes, strengthening threat detection capabilities, and enabling adaptive security measures. This collaboration between advanced technologies contributes to more resilient and proactive defense mechanisms against evolving cyber threats in data transmission.

12.7 CHALLENGES AND FUTURE DIRECTIONS

Securing information in transit and leveraging AI/ML for robust data protection present both challenges and promising future directions in cybersecurity. Here's an exploration of these aspects [7] [8]:

Challenges:

Adaptation to Evolving Threats: Cyber threats are constantly evolving, presenting a challenge for AI/ML-based security systems to keep pace. Attackers continually devise new methods to bypass encryption, requiring AI/ML models to swiftly adapt and counter emerging threats.

Complexity of AI-Driven Security: Implementing AI/ML-powered security solutions requires specialized expertise. Organizations face challenges in deploying and managing sophisticated AI algorithms, necessitating skilled professionals for effective implementation and maintenance.

Privacy Concerns and Ethical Considerations: AI/ML systems analyzing sensitive data for threat detection raise concerns about privacy breaches and ethical implications. Balancing effective threat detection with user privacy and ethical considerations remains a challenge.

Integration and Compatibility Issues: Integrating AI/ML technologies with existing security infrastructure, such as SSL/TLS, poses integration challenges. Compatibility issues, data interoperability, and ensuring seamless integration without disrupting existing systems are hurdles to overcome.

Robustness of AI-Driven Security: AI models are susceptible to adversarial attacks or manipulations. Ensuring the robustness and reliability of AI/ML algorithms in detecting threats and avoiding false positives/negatives remains a challenge.

Future Directions:

AI/ML-Driven Threat Intelligence: Advancements in AI/ML will focus on leveraging big data and analytics for more sophisticated threat intelligence. This includes predicting future threats, proactively identifying vulnerabilities, and enhancing real-time threat detection.

Autonomous Security Response: Future AI/ML systems will evolve to autonomously respond to threats in real-time, taking immediate action to mitigate risks. Automated incident response mechanisms will become more refined and effective.

Explainable AI for Security: Developing AI models that provide transparent explanations for their decisions will gain importance. 'Explainable AI' ensures the understanding of AI/ML-driven security decisions, aiding in compliance and trust-building.

Quantum-Safe Encryption Integration: With the evolution of quantum computing, the focus will shift towards integrating quantum-safe encryption methods with AI/ML-driven security systems, ensuring resilience against quantum-based threats.

Collaborative Security Ecosystems: Future directions will emphasize collaborative security ecosystems. Interoperability among AI/ML-based security solutions, standardization of protocols, and sharing threat intelligence for collective defense will become vital.

Securing information in transit by leveraging AI/ML for robust data protection requires addressing current challenges while steering towards future advancements. The focus will be on adaptive, intelligent, and collaborative security solutions capable of defending against dynamic cyber threats in an increasingly interconnected digital landscape.

12.8 BEST PRACTICES AND RECOMMENDATIONS

Utilize strong encryption protocols like TLS/SSL to encrypt data during transmission. Leverage AI/ML to enhance encryption key management, ensuring the generation, distribution, and storage of robust encryption keys. Employ AI/ML-driven monitoring systems to continuously analyze network traffic and user behaviors [9]. Detect anomalies or deviations from normal patterns in real-time, enabling prompt response to potential threats. Integrate AI/ML systems with threat intelligence feeds and databases. Leverage real-time threat intelligence to stay updated on emerging threats and adapt security measures accordingly. Use AI/ML for behavioral analysis to identify unusual user behaviors or access attempts during data transmission. Implement adaptive access control measures based on behavior patterns.

Develop AI/ML-powered automated response mechanisms to handle security incidents in real-time. Enable quick isolation of affected systems or initiation of incident response protocols to mitigate risks. Conduct frequent security audits to identify vulnerabilities in AI/ML-driven security systems. Ensure timely updates and patches to maintain the effectiveness and resilience of security measures. Implement privacy-preserving AI/ML techniques to ensure user privacy while analyzing sensitive data for threat detection. Strive for a balance between effective security measures and user privacy. Use explainable AI techniques to ensure transparency in AI/ML-driven security decisions. Provide understandable explanations for security decisions to build trust and facilitate compliance. Foster collaboration among security teams and share threat intelligence within and across organizations. Encourage collective defense strategies and information sharing to combat evolving threats. Anticipate the future impact of quantum computing on encryption. Start exploring and preparing for quantum-safe encryption methods to ensure resilience against future quantum-based threats.

By following these best practices and leveraging AI/ML technologies, organizations can bolster their defenses, detect threats more effectively, and ensure robust protection for data during transmission. This approach aligns with evolving security needs, providing a proactive stance against dynamic cyber threats in today's interconnected digital environment.

12.9 CASE STUDIES AND USE CASES

A few case studies and use cases that highlight the application of AI/ML for securing information in transit and ensuring robust data protection [10] [11]:

Cisco Umbrella's AI-Driven Threat Defense:

Case Study: Cisco Umbrella uses AI/ML algorithms to analyze global internet activity, identifying malicious domains and threats in real-time. Use Case: AI/ML models continuously monitor DNS requests and patterns, detecting anomalies indicative of potential threats. It blocks malicious domains, preventing users from accessing harmful sites even before a connection is established. Darktrace's AI Cyber Defense:

Case Study: Darktrace employs AI-powered anomaly detection to secure information in transit across networks. Use Case: Darktrace's AI learns 'normal' behaviors within networks and identifies deviations, flagging potential threats during data transmission. It recognizes anomalous traffic patterns or unusual data transfers, allowing proactive threat response. Google's TensorFlow for Network Security:

Case Study: Google utilizes TensorFlow, an open-source ML platform, for network security and threat detection. Use Case: TensorFlow-based models analyze network traffic patterns, identifying potential threats or malicious activities during data transmission. It assists in predicting and mitigating cyber threats, enabling faster response times.

IBM Security QRadar SIEM:

Case Study: IBM Security QRadar SIEM integrates AI/ML for threat detection and response [12]. Use Case: AI-powered analytics within QRadar SIEM analyze network logs, traffic, and user behaviors. It detects suspicious activities or anomalies in data transmission, correlating information across diverse sources for comprehensive threat visibility and automated incident response. Cylance's ML-Driven Endpoint Protection:

Case Study: Cylance employs ML for endpoint protection, securing data on devices during transmission. Use Case: ML models on endpoints analyze file structures and behaviors, identifying and blocking potential threats during data transfer. It prevents malware execution or data breaches by detecting malicious files or activities in transit.

These case studies and use cases showcase the practical implementation of AI/ML technologies in securing information during transmission. AI/ML-driven solutions play a pivotal role in threat detection, anomaly identification, and proactive defense mechanisms, ensuring robust data protection across various networks and systems [13] [14].

12.10 CONCLUSION

Securing information in transit through the utilization of Artificial Intelligence (AI) and Machine Learning (ML) stands as a crucial strategy in fortifying data protection in today's interconnected digital landscape. AI/ML technologies offer advanced capabilities that redefine the approach to safeguarding data during transmission across networks or channels. The integration of AI/ML in data protection measures presents a proactive and adaptive defense against evolving cyber threats. By continuously analyzing patterns, behaviors, and anomalies in real-time, AI/ML-powered systems enable the early detection of potential risks, ensuring swift response and mitigation.

AI/ML's role extends beyond traditional encryption methods; it encompasses predictive analytics, behavioral analysis, threat intelligence integration, and automated response mechanisms. These technologies enable adaptive security measures, ensuring resilience against a dynamic threat landscape. However, challenges persist, including the need for skilled expertise, privacy concerns, and the ever-evolving nature of cyber threats. Balancing effective threat detection with user privacy and ethical considerations remains an ongoing concern. Looking ahead, the future of securing information in transit relies on collaborative ecosystems [15], quantum-safe encryption advancements, explainable AI, and the seamless integration of AI/ML-driven security solutions. The emphasis lies on adaptive, intelligent, and transparent systems that address emerging threats while upholding user privacy and ethical standards.

In conclusion, leveraging AI/ML for robust data protection during information transmission is pivotal in establishing a proactive defense posture. It's a continuous journey that demands a multidimensional approach, collaboration, and ongoing innovation to counteract the ever-evolving threat landscape and ensure the integrity and confidentiality of data in transit.

REFERENCES

1. Singh, J., 2014. Real Time BIG Data Analytic: Security Concern and Challenges with Machine Learning Algorithm. 2014 Conference on IT in Business, Industry and Government (CSIBIG), pp. 1–4. https://doi.org/10.1109/CSIBIG.2014.7056985
2. Nandakumar, K., Vinod, V., Batcha, S., Sharma, D., Elangovan, M., Poonia, A., Basavaraju, S., Dogiwal, S., Dadheech, P., Sengan, S., 2021. Securing Data in Transit Using Data-in-Transit Defender Architecture for Cloud Communication. *Soft Computing*, 25, pp. 12343–12356. https://doi.org/10.1007/s00500-021-05928-6
3. Liu, Z., 2021. Construction of Computer Mega Data Security Technology Plat-form Based on Machine Learning. *2021 IEEE 4th International Conference on Information Systems and Computer Aided Education (ICISCAE)*, pp. 538–541. https://doi.org/10.1109/ICISCAE52414.2021.9590732
4. Batham, D., 2022. Information Safety in Cloud Computing. *Praxis International Journal of Social Science and Literature*. 6, pp. 4–8. https://doi.org/10.51879/pijssl/050601

5. Albugmi, A., Alassafi, M., Walters, R., Wills, G., 2016. Data Security in Cloud Computing. *2016 Fifth International Conference on Future Generation Communication Technologies (FGCT)*, pp. 55–59. https://doi.org/10.1109/FGCT.2016.7605062
6. Sayadi, H., Aliasgari, M., Aydin, F., Potluri, S., Aysu, A., Edmonds, J., Tehranipoor, S., 2022. Towards AI-Enabled Hardware Security: Challenges and Opportunities. *2022 IEEE 28th International Symposium on On-Line Testing and Robust System Design (IOLTS)*, pp. 1–10. https://doi.org/10.1109/IOLTS56730.2022.9897507
7. Bhardwaj, A., Kaushik, K., 2022. Predictive Analytics-Based Cybersecurity Framework for Cloud Infrastructure. *International Journal of Cloud Applications and Computing*. 10, pp. 10–15. https://doi.org/10.4018/ijcac.297106
8. Bhardwaj, A., Kaushik, K., 2022. Predictive Analytics-Based Cybersecurity Framework for Cloud Infrastructure. *International Journal of Cloud Applications and Computing*. 12, pp. 100–106. https://doi.org/10.4018/ijcac.297106
9. Bock, H., Morais, R., Pedro, J., Sommerkorn-Krombholz, B., Sadasivarao, A., Syed, S., Paraschis, L., Kandappan, P., 2020. Coming of Age of AI-Assisted Network Management Control. . https://doi.org/10.1364/networks.2020.new1b.2.
10. Bhardwaj, A., 2019. Solutions for Securing End User Data over the Cloud Deployed Applications. *Cloud Security*. 12, pp. 60–68. https://doi.org/10.4018/978-1-5225-8176-5.ch053.
11. Heni, H., Gargouri, F., 2015. A Methodological Approach for Big Data Security: Application for NoSQL Data Stores., 12, pp. 685–692. https://doi.org/10.1007/978-3-319-26561-2_80
12. Dhingra, M., 2015. Cloud Data Encryption Ensuring Security. *International Journal of Engineering Research and Technology*, 4, pp. 60–66.
13. Poonguzhali, E., SuhasRao, M., Gk, S., Khanum, M., 2018. Protection and Security of Data in Cloud Computing. *International Journal of Engineering Research and Technology*, 5, pp. 45–50.
14. Alaoui, I., Gahi, Y., 2020. Network Security Strategies in Big Data Context., pp. 730–736. https://doi.org/10.1016/j.procs.2020.07.108
15. Raj, P., Mahmood, Z., 2021. Citizen Data in Distributed Computing Environments., pp. 183–203. https://doi.org/10.4018/978-1-7998-4570-6.ch009.

13 Optimizing Medical Image Compression for Efficient Information System Integration
A Comprehensive Review

Kanwaldeep Kaur Sidhu, Heena Talat, and
Intisar S. Al-Mejibli

13.1 INTRODUCTION

The optimization of medical image compression is essential for smooth integration with information systems in the quickly changing field of healthcare technology. This will transform the way physicians access and use diagnostic imaging data. In-depth analysis of the complex interactions between information system integration and medical picture reduction techniques is provided in this thorough review, which also examines tactics to improve productivity, accessibility, and interoperability in healthcare settings.

Fax machines with digital technology, like the Bartlane Cable Picture Transmission System, were developed decades before digital cameras and computers. The Standards Eastern Automatic Computer (SEAC) at NIST presented the first image to be examined, saved, and replicated in digital pixels [1]. Beginning in the early 1960s, digital imaging continued to progress along with the medical research and space program. Projects at Bell Labs, University of Maryland, MIT, Jet Propulsion Laboratory, and Bell Labs used digital images, among other places, to improve medical imaging, satellite imagery, character recognition, videophone technology and photo enhancement [2]. Microprocessors, along with related advancements in storage and display technology, were introduced in the early 1970s and led to rapid developments in digital imaging. By the end of the 20th century, analog film and tape were gradually replaced by charge-coupled devices (CCDs), which were developed and marketed for use in a variety of image-capturing devices thanks to advancements in microprocessor technology. The processing power required to handle digital picture capture also made it possible for computer-generated digital images to approach photorealism in terms of refinement [3]. Images might be handled in real time for some specific issues, such as converting television standards. Among image processing techniques, digital image processing (DIP) has been the most widely used. This method is usually chosen since this is the least expensive and most adaptable.

13.2 DESCRIPTION OF IMAGE AND DIGITAL IMAGE

An image is a surface-produced visual representation of an item, scene, person, or abstraction. The word image comes from the Latin word *imago*. A sequence of images shown so quickly that the eye can integrate them is called a video. Images can be two-dimensional (like a picture or a screen projection) or three-dimensional (like a statue or hologram). Optical tools like cameras, telescopes, microscopes, mirrors, and the like, as well as natural events and things like the human eye and water surfaces, can catch them. In practically every subject, including the sciences, mathematics, and medicine, images are essential. An example of a two-dimensional functional for a digital image is $f(p, q)$, where p and q are spatial coordinates and Any pair of coordinates has an amplitude, denoted by f of p, q that represents the intensity or gray level of the image at that particular position. Digital pictures are computer photos that have been scanned from materials, such as artwork, written texts, manuscripts, and photos.

A two-dimensional image represented numerically is called a digital image. Digital images can be either vector or raster-based. Pixels, or picture elements, make up raster-type digital images. These pixels are kept in computer memory as raster images or raster maps, which are two-dimensional arrays of tiny numbers. At a single point in the image, every pixel is given a tonal value, such as white, black, grayscale, or color, which is indicated in binary code as ones and zeros. Pixels are organized in consistent patterns of rows and columns and information is stored differently from grain particles in a traditional photographic image. The computer stores every pixel's binary digits/bits in a series and often reduces them in a mathematical representation (compressed form). The computer then reads and interprets the bits to create an analog version that can be printed or displayed. Digital equipment is used to take, save, edit, and view photographic images. Prior to use, the images must be digitized, or scanned, into a set of numbers. Once an image has been digitalized you can use the computer to preserve, inspect, change, show, communicate, or print the photographs in an amazing variety of methods since computers are exceptionally good at manipulating and storing numbers.

Images in the vector style were produced by mathematical vectors. To put it simply, a vector is made up of points with both length and direction. An image's resolution is defined as its pixel density. An image holds more information the greater its resolution. An alternative way to quantify image resolution is as the pixel array's dimension, which may be expressed as two numbers: total number of pixels/1000000 then round of it [4].

13.2.1 Digital Image Types

There are two major categories of digital photographs for the purpose of photography purposes: colored, and black and white. Black and white images are composed of grayscale shade pixels, and colored pixels make up colored visuals.

13.2.2.1 Grayscale Images

Traditionally, a grayscale image consists of 8 bpp (bits-per-pixel) pixels, every one of which contains a single value that represents the image's gray level at that specific position. These gray levels cover the entire spectrum in a series of extremely small steps, often consisting of 256 distinct gray colors, from 0 (black) to 255 (white). Given that the human eye can only discern roughly 200 distinct shades of gray.

13.2.2.2 Color Images

Each pixel in a digital colored image is represented by 24 bits (3 bytes) and is stored in a three-dimensional array. Three integers are stored in each pixel, which represent the image's red, green, and blue levels at a specific spot. RGB format, or red, blue and green color component scale, spans 0 to 255. The fundamental colors RGB, often known as additive primary colors, are used to combine light. Red, green, and blue light can be combined in the right proportions to generate any color. This translates to around 16.7 million distinct color combinations. Keep in mind that the white and black version will require three times less memory than a colored one for photographs of the same size.

13.2.2.3 Binary Images

In a binary image, a single bit represents every pixel. Because a bit is limited to two states, such as on or off, each pixel in the binary image should be one of two colors, usually white or black. Their utility in handling photographic images is limited by their incapacity to depict intermediate shades of gray.

13.2.2.4 Indexed Color Images

Certain colored graphics are produced using a small color palette—usually 256 distinct colors. Because each pixel in these photographs has a palette index that indicates which color in the palette corresponds to that pixel, the images are called indexed color images. Representing photographic images with indexed color has a number of drawbacks. First, the image is degraded if methods like dithering are used to depict colors that are absent from the palette since there are more distinct colors in the image than there are in the palette. Second, the limited amount of available colors also causes issues when integrating two indexed color images that employ distinct palettes.

13.2.2 Digital Image Processing with Information System Optimization

This investigation explores the intersection of information system optimization and digital image processing techniques, shedding light on the possible benefits and difficulties of combining these technologies in the healthcare industry. DIP is the process of modifying digital photos with a digital computer. It is a sub-field of signals and systems that focuses specifically on the images. Building an image-processing computer system is the main goal of DIP and the input to the system is the digital image. It processes the image using effective algorithms and outputs an image as the result. The most often used illustration is Adobe Photoshop. It is one of the most popular programs for manipulating digital photos. Compared to analog image processing, digital image processing offers numerous benefits. For example, a greater variety of algorithms can be used on the input data, and processing issues like signal distortion and noise can be avoided. Images are defined in both two and many dimensions. The two primary objectives of digital image processing are to improve visual information for human interpretation and to process the image data for storage, transfer, and description for autonomous machine perception.

13.2.3 Basic Digital Image Processing Steps

The initial phase in the procedure is image acquisition. It might just be as easy as receiving an already-digitized photograph. Typically, pre-processing is done during the picture capture phase. Image enhancement is the easiest and most captivating use of digital image processing. A highly individualized field of image processing is picture enhancement. Enhancement techniques, such as adjusting brightness and contrast, are essentially used to highlight specific aspects of interest in an image or to bring out detail that is hidden. The goal of picture restoration is to improve an image's appearance. Nonetheless, picture restoration is objective since most restoration methods are predicated on probabilistic or statistical models of image deterioration. Due to the ensuing rise in the usage of digital images on the Internet, color image processing has grown in significance. Among other things, this could include digital color modeling and processing.

The basis for displaying images at different resolutions is wavelets. This is specifically utilized for pyramidal representation—a technique that divides images into progressively smaller regions—and image data compression. Compression techniques deal with ways to lower the amount of bandwidth needed to transmit a picture or the amount of storage space needed to save it.

Most computer users are probably aware of image compression (maybe unintentionally) thanks to image file extensions. Morphological processing focuses on tools for extracting picture components useful for form representation and description. Segmentation techniques separate the objects that make up an image. Generally speaking, distinct segmentation is one of the most challenging issues in DIP. When it comes to handling image issues where individual object recognition is required, a strong segmentation technique is essential. The illustration and detail come after the result of a segmentation phase, which is typically raw pixel data comprising all the points within a location or its boundary. Choosing a representation is only one stage in the process of transforming raw data into a format that computers can process in the future. Description is the process of extracting characteristics that either provide quantitative information of interest or are required to differentiate one class of objects from another. The process of labelling an object based on its descriptors is called object recognition. To reduce the amount of search work required to find the desired information, knowledge can be as basic as identifying areas of a picture where the information is known to be present. A knowledge base having satellite image with high resolution photos of an area in relation to change-detection applications, or an image database including an interconnected list of all significant potential faults in a materials inspection problem, are two examples of how complicated a knowledge base may be. The way the modules interact with one another is likewise managed by the knowledge base.

13.3 MEDICAL IMAGING WITH INFORMATION SYSTEMS

Many aspects of human existence are becoming digitized these days. A significant amount of digital information must be stored for this. All of this is made possible by the advancement in the registration of various types of data. This general advancement is shown in the large field of digital images, which includes photos used in medicine. The smooth integration of medical imaging with information systems in the digital era of healthcare is transforming clinical workflows, facilitating real-time collaboration, and encouraging data-driven decision-making for better patient outcomes. Following Professor Roentgen's discovery of X-rays in the first decade of the 1900s, radiology emerged as a subspecialty of medicine. Up until World War II, radiology development advanced quickly. Over the past 25 years, there has been an explosion in the development of diagnostic imaging techniques due to the widespread use of X-ray pictures during World War II, the introduction of digital computers, and the introduction of novel imaging modalities including magnetic resonance imaging and ultrasound.

Digital image processing technology for medical applications was achieved in the Space Foundation Space Technology in 1994. The diagnosis of illnesses and surgical planning are significantly impacted by medical imaging [5]. Medical image is the process, methodology, and the practice of producing visual representations inside the body for therapeutic intervention and clinical assessment. Medical imaging uses images of internal organs that are concealed by the skin and bones can be diagnosed and cure diseases. Healthcare imaging also builds a repository for typical physiology and anatomy to identify anomalies. Medical imaging can be performed on removed organs and tissues, but this type of procedure is usually categorized as pathology rather than medical imaging. Novel techniques and applications, including medical photography techniques, elastography, tactile imaging, CT, MRI, endoscopy, and thermography—are created as health care becomes more computerized. An object's cross-section is produced by image stacks, which are sequences of pictures produced by MR and CT. Doctors at hospitals maintain extensive records of medical examination results, enabling them to diagnose patients based on a variety of examination findings. Patients can obtain valuable information about their medical records through these platforms. The IZIP-Czech system, which provides online access to patient health records, is a really good example.

Hospital databases are growing quickly every day, and thousands of photos must be recorded each day for a certain amount of time so that the data can be used later. One needs to maintain good quality data, short transmission times, and little storage space when replacing image data. Unfortunately, the increase in throughput of utilized communication links is not adequate to meet needs; therefore, further solutions need to be pursued. Costs go up with increased capacity as well. Therefore, research and development are being done on picture compression approaches to address these issues. Image compression can be approached from a variety of angles, and each angle results in a wide range of techniques. Medical photos need to be compressed with extra care because poor-quality images can be hazardous to patients' health and create distortion, which can affect the accuracy of diagnosis. At first, information preservation techniques were the main focus. Takaya et al. conducted research on scan pixel disparity in [6]. In [7], Assche et al. take advantage of the inter-frame redundancy. In [8], linear predictive coding techniques were examined. Worldwide, 5 billion medical imaging investigations have been carried out as of 2010 [9]. In the United States in 2006, the radiation exposure resulting from medical imaging accounted for approximately 50% of all ionizing radiation exposure [10].

13.3.1 Procedures for Medical Images

As direct digital imaging solutions for medical diagnostics become more and more in need, DIP plays an increasingly significant role. The National Electrical Manufacturers Association (NEMA) created the Digital Imaging & Communications in Medicine (DICOM) standard to make this process easier [11]. This standard facilitates the evaluation and distribution of medical imaging tests, such as ultrasounds, CT scans, and MRIs. Diagram (Fig. 13.1) illustrates the four main domains of DIP for medical image processing [12].

1. Image Formation: This covers the entire process, from taking the picture to laying up the digital image matrix. Forming includes digitalization and acquisition.
2. Image Enhancement: This involves every stage of picture transformation, registration, and filtering.
3. Image Visualization: It lowers all manipulation classes in order to get the best possible image output from that matrix. It includes processes for lighting, shading, and displaying.
4. Image Analysis: This covers every stage of image processing, such as segmentation, classification, and feature extraction. Prior knowledge of the system's nature and content is required for these steps.
5. Image Management: It provides an overview of all the methods for efficiently storing, transmitting, archiving, and retrieving image data.

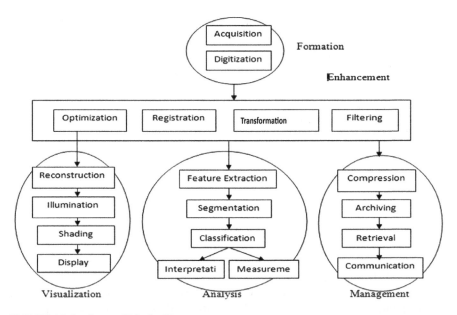

FIGURE 13.1 Steps of Medical Image.

From image acquisition and interpretation to treatment planning and follow-up care, the integration of medical imaging with information systems provides a unified platform for managing the full patient care continuum, advancing precision medicine and improving global medical services. [13] proposed a framework for shadow preservation enhancing the content-aware image retargeting process, contributing to improved image quality and visual appeal. The proposed framework, verified by means of numerical assessments and graphical comparisons, demonstrates its efficacy in advancing content-aware image retargeting applications.

13.4 INTEGRATION OF MEDICAL IMAGE COMPRESSION WITH INFORMATION SYSTEMS

An important development in healthcare data management is the use of image compression methods in information systems. These integrated solutions maximize medical image compression while maintaining diagnostic quality during storage, transmission, and retrieval. Because of this synergy, electronic health record (EHR) systems can store massive volumes of imaging data more effectively, allowing for easy access and analysis. Additionally, the expedited transfer of condensed images facilitates telemedicine programs and improves communication among medical practitioners, especially in remote or resource-constrained settings. Consequently, the incorporation of image compression into information systems raises the standard of patient care by increasing the accessibility and efficacy of medical imaging data. For purposes like data transmission and storage, image compression is essential. Reducing duplication and irrelevance of the image data is the primary goal of the image [14]. The goal of the method is to produce a condensed image demonstration, which will lower the amount of storage space and transmission rate.

It strives for a high compression ratio while maintaining an image that is nearly identical in terms of quality and resolution to the original. To achieve compression, one or three fundamental data redundancies must be eliminated:

1. Coding redundancy
2. Interpixel redundancy
3. Psychovisual redundancy

When it comes to code redundancy, a code is simply a collection of symbols, and each code word is a sequence of symbols that conveys a piece of information. The length of a code word (8 bits) indicates how many symbols are in each code word. It appears when less-than-ideal code phrases are used. Correlated pixels lead to interpixel redundancy. when the sensitivity with which the human eye responds to different types of visual information, psychovisual redundancy occurs.

13.4.1 Need of Image Compression

Multimedia (music, video, and graphics) data that is not compressed needs a large amount of storage space and transmission bandwidth. Although processing speeds, mass storage density, and the effectiveness of digital communication systems have advanced at a rapid pace, there is still a significant need for data transmission bandwidth and data storage capacity. The need for more efficient picture and signal encoding techniques has not only increased in recent years due to the rise in data-intensive multimedia-based web applications, but it has also made signal compression essential to communication and storage technologies.

- It provides cost savings as less data is sent over a switched telephone network.
- It also shortens execution times and the amount of storage space needed.
- It minimizes the chances of transmission errors as less data is transferred.
- I also provides a security level against illegitimate monitoring.

13.4.2 Usage of Image Compression:

- Satellite imaging for remote sensing.
- Medical imaging and documentation.
- Transmission using facsimile (FAX).
- Command of remotely operated vehicles for space exploration, the military, and hazardous waste disposal.

13.5 METHODS FOR MEDICAL IMAGE COMPRESSION

There are mainly two methods of image compression: lossless and lossy, depending on whether an exact model is completely recreated.

- *Lossless Compression*

When an image is compressed using lossless compression, the reconstructed image has the same dimensions as the original. It can only compress data to a certain degree. Since it doesn't introduce noise to the signal or image, it is also known as noiseless. Only a few applications, including medical imaging, where information loss shouldn't occur, use this kind of compression. Entropy coding is another name for lossless coding [15]. PNG and GIF are two types of lossless [16] image compression.

- *Run Length Encoding (RLE)*

RLE is a very basic compression technique that is applied to data that is sequential [17]. Additionally, it supports data that is repeated. It substitutes the so-called runs—sequences of comparable pixels. For grayscale images, the sequence {Vi, Ri} represents the run length coding; Vi is the pixel intensity, wherein Ri denotes the quantity of succeeding pixels inside an intensity. Fax machines are the main use for RLE.

- *Huffman Encoding*

Huffman is the one who invented this encoding. It is currently utilized as a backend for various other compression techniques [18]. With this method, symbols were coded according to their occurrence probabilities or frequencies using a variable length code table. The image's pixels are regarded as symbols. A larger number of bits is assigned to less frequent pixels and a smaller number of bits is allotted to more frequent pixels. In the majority of compression systems, Huffman coding serves as the last step as a lossless compression approach following the use of lossy compression techniques.

- *Arithmetic Coding*

One type of entropy coding is arithmetic coding. In most cases, Each character in a string of characters like "hello there" is represented by a certain number of bits, like in ASCII coding. Characters that are used frequently will be saved with tiny bits when a string is converted to arithmetic encoding, using fewer bits overall [19]. Arithmetic encoding reduces the message's whole length to a single number, in contrast to other entropy encoding techniques like Huffman coding [20], which divides the input into its component symbols and adds a code to each one.

- *Area Coding*

An enhanced version of RLE that captures the two-dimensional nature of images is area coding. It is constructed with two-dimensional items and features a variety of sequences. Rectangular regions with similar properties are found using the area coding algorithm. These areas are represented as an element with two points in an expressive code. Although this coding has a nonlinear technique, it can nevertheless be quite successful [17].

- *LZW Coding*
Lempel, Ziv, and Welch developed the Lempel–Ziv–Welch (LZW) algorithm, which is a worldwide lossless data compression method. As an enhanced version of the LZ78 algorithm released by Lempel and Ziv in 1978, it was released by Welch in 1984 [18]. This type of coding is dictionary-based and can be either static or dynamic. In dynamic, the dictionary is updated dynamically; in static, it is fixed during the encoding and decoding process.

- *Lossy Compression*
The reconstructed image with lossy compression is roughly the original image's size, but it is not exactly the same [21]. Additionally, when the redundancy in the signal or image is eliminated entirely, the image is degraded. The quantization technique, which divides the data into distinct bins and assigns a value to each bin, also results in information loss but has a substantially higher compression ratio.

- *Fractal Coding*
The primary goal of this is to use common image processing methods like edge detection, color separation and texture analysis to break the image up into smaller pieces, or segments. Next, every segment was looked up in a fractal library. Iterated function system (IFS) codes, which make up the library, are essentially collections of codes. A collection of codes for a certain image is found by following the correct approach. This coding system is quite useful for compressing uniform and well-self-similar images.

- *Vector Quantization*
A traditional quantization method from signal processing called vector quantization (VQ) enables the distribution of prototype vectors to be used as a model for probability density functions. The creation of a dictionary of fixed-size vectors, or "code vectors," is the fundamental principle behind this technique. A block of pixel values typically makes up a vector. Next, a given image is divided into image vectors, which are non-overlapping blocks. Next, each entry in the dictionary is identified, and the original image vector is encoded using that entry's index. Consequently, a series of indices that can be further entropy coded serve as the representation for each image. [22].

- *Block Truncation Coding*
For this coding, the image needs to be split up into non-overlapping pixel blocks. The reconstruction and threshold values are computed for every pixel. The threshold in this instance is frequently the mean of the pixel values. After that, a bitmap is created by changing each pixel in the block whose values are larger than or equal to the threshold. Next, each segment's reconstruction value is ascertained.

- *Sub-bands Coding*
The image is analyzed in this coding process to create the components that contain frequencies in distinct bands, or sub bands. As a result, quantization and coding are performed to each sub band, which is appropriate because each sub band can be constructed independently.

13.6 ROI AND REGION GROWING IN MEDICAL IMAGING

Region growth and ROI selection are frequently combined in medical imaging applications. For instance, a ROI may be first constructed manually or automatically. Then, region growth may be used inside the ROI to further segment or refine particular abnormalities or structures inside the chosen area. In tasks like image analysis, feature extraction, and computer-aided diagnosis in medical imaging, both approaches are crucial. They make it possible for medical professionals and researchers to concentrate on pertinent regions of images and derive important data for study, treatment planning, and diagnosis.

13.6.1 Region of Interest (ROI)

The limits of lossy and lossless compression algorithms are taken into consideration when designing ROI. With lossy coding methods, the compression ratio can reach up to 5–30%, whereas most lossless compression techniques have a compression ratio of about 80% of the original size, [6], [8], although there can be a significant data loss. Because medical images do not allow for any information loss in crucial areas for diagnosis, ROI is primarily introduced for these types of images. A medical image is separated into ROI and non-ROI components. The ROI portion—also referred to as the foreground portion—is thought to be the most crucial diagnostic component, while the remaining portion is referred to as the non-ROI portion, or background portion. Therefore, it is necessary to use the lossless compression technique in order to maintain the diagnostic part's quality (ROI) and to offer a high compression ratio [6], [23]. The ROI portion of the image must be communicated first or with a higher priority during image transmission for telemedicine purposes, meaning that the ROI-related coefficients are conveyed before the non-ROI-related coefficients.

13.6.2 Region Growing

One of the easiest area-based techniques is region growth. Another name for it is the "pixel-based method." The region-growing approach looks at all of the neighboring pixels around the main seed points to see if they are contributing to the region or not. This process is identical to the data clustering technique. First, a selection of seeds is made for region growing. The user determines the criterion for selecting the seed point, which may include pixel intensity, texture, color level, or grayscale. The primary region begins exactly where the seed point is, and the seed point (chosen pixel) chooses all of the surrounding pixels to include in the region. In the expanding region, picture data is equally crucial.

13.7 CONCLUSION

To sum up, combining information system optimization with medical image reduction is a big step in the right direction for improving healthcare delivery in the digital era. Healthcare organizations may efficiently handle the increasing amount of medical imaging data while guaranteeing effective storage, transmission, and retrieval operations by enhancing compression strategies. Furthermore, the smooth incorporation of compressed images into information systems enhances data-driven decision-making, interdisciplinary cooperation, and clinical workflow optimization.

Future developments in information system optimization and medical image compression hold the potential to significantly alter the way healthcare is provided. The capacities of information systems and compression algorithms can be improved through the use of emerging technologies like artificial intelligence and cloud computing, which will ultimately advance precision medicine and individualized patient care.

REFERENCES

1. Rosenfeld, Azriel. "Picture processing by computer." *ACM Computing Surveys (CSUR)* 1, no. 3 (1969): 147–176.
2. Gonzalez, Rafael C. *Digital image processing*. Pearson Education India, 2009.
3. Jähne, Bernd, ed. *Spatio-temporal image processing: Theory and scientific applications*. Berlin, Heidelberg: Springer Berlin Heidelberg, 1993.
4. Dallwitz, M. J. "An introduction to computer images." *TDWG Newsletter* 7, no. 10 (1992): 1–3.
5. Brindha, B., and G. Raghuraman. "Region based lossless compression for digital images in telemedicine application." In *2013 International Conference on Communication and Signal Processing*, pp. 537–540. IEEE, 2013.
6. Miaou, Shaou-Gang, Fu-Sheng Ke, and Shu-Ching Chen. "A lossless compression method for medical image sequences using JPEG-LS and interframe coding." *IEEE Transactions on Information Technology in Biomedicine* 13, no. 5 (2009): 818–821.
7. Placidi, Giuseppe. "Adaptive compression algorithm from projections: Application on medical greyscale images." *Computers in Biology and Medicine* 39, no. 11 (2009): 993–999.
8. Baeza, Ismael, J.-A. Verdoy, Javier Villanueva-Oller, and R.-J. Villanueva. "ROI-based procedures for progressive transmission of digital images: A comparison." *Mathematical and Computer Modelling* 50, no. 5–6 (2009): 849–859.
9. Roobottom, C. A., G. Mitchell, and G. Morgan-Hughes. "Radiation-reduction strategies in cardiac computed tomographic angiography." *Clinical Radiology* 65, no. 11 (2010): 859–867.
10. Mettler Jr, Fred A., Mythreyi Bhargavan, Keith Faulkner, Debbie B. Gilley, Joel E. Gray, Geoffrey S. Ibbott, Jill A. Lipoti et al. "Radiologic and nuclear medicine studies in the United States and worldwide: frequency, radiation dose, and comparison with other radiation sources—1950-2007." *Radiology* 253, no. 2 (2009): 520–531".
11. Shah, Rushabh, Priyanka Sharma, and Rutvi Shah. "Performance analysis of region of interest based compression method for medical images." In *2014 Fourth International Conference on Advanced Computing & Communication Technologies*, pp. 53–58. IEEE, 2014.
12. Deserno, Thomas Martin. "Biomedical Image Processing (Biological and Medical Physics, Biomedical Engineering)." (2001).
13. Grasemann, Uli, and Risto Miikkulainen. "Effective image compression using evolved wavelets." In *Proceedings of the 7th annual conference on Genetic and evolutionary computation*, pp. 1961–1968. 2005.

14. Garg, Ankit, Aleem Ali, and Puneet Kumar. "Original Research Article A shadow preservation framework for effective content-aware image retargeting process." *Journal of Autonomous Intelligence* 6, no. 3 (2023): 1–20.
15. Jain, Anil K. "Image data compression: A review." *Proceedings of the IEEE* 69, no. 3 (1981): 349–389.
16. Yang, Ming, and Nikolaos Bourbakis. "An overview of lossless digital image compression techniques." In *48th Midwest Symposium on Circuits and Systems, 2005*, pp. 1099–1102. IEEE, 2005.
17. Sonal, Dinesh Kumar. "A study of various image compression techniques." *COIT, RIMT-IET. Hisar* 8 (2007): 97–102.
18. Mathur, Mridul Kumar, Seema Loonker, and Dheeraj Saxena. "Lossless Huffman coding technique for image compression and reconstruction using binary trees." *International Journal of Computer Technology and Applications* 3, no. 1 (2012).
19. Vijayvargiya, Gaurav, Sanjay Silakari, and Rajeev Pandey. "A survey: various techniques of image compression." arXiv preprint arXiv:1311.6877 (2013).
20. Jagadish, H. Pujar, and M. Kadlaskar Lohit. "A new lossless method of image compression and decompression using Huffman coding techniques." *Journal of Theoretical and Applied Information Technology* 12, no. 2 (2010): 18–23.
21. Parmar, Chandresh K., and Kruti Pancholi. "A Review on Image Compression Techniques." *Journal of Information. Knowledge and Research in Electrical Engineering* 2, no. 2 (2015): 281–284.
22. Chen, Wen-Shiung, En-Hui Yang, and Zhen Zhang. "A new efficient image compression technique with index-matching vector quantization." *IEEE Transactions on Consumer Electronics* 43, no. 2 (1997): 173–182.
23. Maglogiannis, Ilias, Charalampos Doukas, George Kormentzas, and Thomas Pliakas. "Wavelet-based compression with ROI coding support for mobile access to DICOM images over heterogeneous radio networks." *IEEE Transactions on Information Technology in Biomedicine* 13, no. 4 (2009): 458–466.

# 14 FaceTrack

A Face Recognition-based Real-time Attendance Marking Approach using Haar Cascade and Machine Learning

R. Rafeek, V. S. Anoop, and Zahid Akhtar

14.1 INTRODUCTION

In today's growing scenario of technological changes and complexities, it has become necessary for organizations to employ the latest technologies to work efficiently and effectively (Gururaj et al., 2024). Of these, a viable system for attendance management emerges as the most needful one as organizations grow and the workforce is geographically distributed (Nguyen et al., 2024). The historical analysis of attendance management shows that the use of such tools passes through different stages starting from simple paper-based tools to advanced technologies. In the early days of industrialization, employees' attendance was recorded very loosely, and it was the practice of foremen to jot down the present workers (Feroze et al., 2024). Time record keeping was first advanced by Willard Bundy who invented the mechanical time clock in 1888 and was succeeded in the middle of the twentieth century by punch card systems that, despite being more efficient than handwritten logs, required a lot of hard work. The centuries ended with utilizing specific technologies such as electronic magnetic stripe cards and barcoding that decreased the amount of time as well as energy required to clock in and prepare payrolls. Such systems also made it easier to manage data and were more accurate than the earlier processes that were in place (Gururaj et al., 2024). Nevertheless, they were not completely shielded from such prevalent problems as time theft and buddy punching, when one worker signs in for another. At the start of the twenty-first century, concepts such as fingerprint scans and even face recognition were developed and were highly accurate methods for identifying employees and tracking their attendance at work. The method was also secure in comparison to the previous techniques and only the concerned employee was allowed to register his or her attendance, making the process much more effective (Chen et al., 2024).

Pertaining to the latter, advancements in computing technologies specifically in cloud computing and mobile technologies have advanced attendance management systems by providing accurate and up-to-date data, integration and compatibility with other human resource and payroll systems, and modularity to accommodate any organization of any size (Rathore et al., 2024). Most of the attendance solution softwares collect data from a single platform and store the data thereby making it convenient to use cloud storage, especially for organizations that have employees spread all over the country or world. Mobile applications enable an employee to be able to punch in and/or punch out using a smartphone or a tablet, which is suitable in situations where the employee is in a different geographical location (Irawati et al., 2024). These advancements complement the increasing use of remote working and cross-border teams since employees' attendance is not manipulated.

However, current historical development has not allowed the traditional work systems to provide solutions for the complexity of contemporary organizations. Old-school systems of attendance may demand the use of expensive equipment like time clocks, card readers and many others hence has a high cost when planning to install them (Opanasenko et al., 2024). They may also entail data entry at some point or periodic update from central databases, which is also prone to errors and lags. Moreover, the conventional systems seldom provide methods that extract and process live data, which pose a major issue in managing the workforce. When it comes to fast decision-making, the system of after-pending checks that are still common in dynamic environments can slow down work and contain inaccuracies in the number of attendees. Also, it proves the need for modern and complex attendance management systems by the absence or inadequacy of old-school approaches (Ali et al., 2024). Contemporary organizations demand from the attending systems not just reliable and secure solutions but also availability of timely data, interconnectivity with other organizational systems, and flexibility supported by the system's ability to adapt to growth and geographical expansion. The issues above can all be solved by using modern automated attendance management systems that are based on the maximized usage of the latest biometric technologies, cloud computing, and the usage of mobile devices (Shukla et al., 2024). They decrease the administrative tasks, lessen the possibility of mistakes, and allow the necessary level of freedom for widespread and non-centralized teams.

When implemented effectively automated attendance management systems increase organizational efficiency, support legal requirements, and provide understanding authority over employees' behavior. They enhance accuracy of the recorded attendance, and this is crucial when it comes to payroll calculation and compliance with laws on working conditions (Abid et al., 2024). They also track true-time statistics of attendance of the employees which can be critical in the decision-making procedures of the managers. On the same note, scalability of the advanced systems is well implemented to guarantee coherence with organizational growth without compromising on their efficiency, hence meeting the requirement of sustainability in the long run. Thus, the more a business develops, the more it will have to rely on proper and effective attendance management (Dang et al., 2023). New generation attendance tracking solutions are not only biometric devices for capturing the presence of subordinates but key organizational assets that contribute to productivity, security and companies' performance. The change from the conventional types of attendance management to the sophisticated systems that spring up is vital for any organization that wants to remain relevant in this modern and rapidly evolving world, which in turn confirms the importance of innovation in deciding the future of work (Budiman et al., 2023). Further, in the process of organizations' further development, the application of AI and machine learning in attendance management systems becomes critical. In this context, AI makes predictive analysis which helps to predict possible attendance troubles based on past data, for example, risky patterns of truancy. In addition, due to approaches based on machine learning, it is possible to work optimal schedules out in terms of the number of employees required and the periods of the day that commonly experience high degrees of demand (Jha et al., 2023). Besides optimizing operations, this approach also aligns with employees' preferences and helps avoid shortages of staff.

Internet of Things (IoT) technology takes attendees' management even to another level. Smart badges and biometric sensors can also allow the organization to capture attendance data without much interference with the employees' daily functioning. These devices are interconnected, as well as with main control systems, ensuring such features as real-time information update and real-time notifications (Kamil et al., 2023). For instance, the authority can be alerted the moment a registered badge is within the restricted area. Such automation and level of integration makes the system more secure and robust in respect to capturing attendees' data which is essential in security conscious and industries that require a high degree of precision (Kumar et al., 2023).

Security and privacy concerns over data have also led to further improvements in attendance data handling. With regulations such as GDPR in Europe, ensuring privacy in handling employee data has attained the highest priority (Chen et al., 2024). Data privacy protection in the latest attendance systems features advanced encryption methods and enhances secure solutions in storage. Needless to say, most such modern attendance systems allow for the provision of configurable privacy settings, which help companies work ethically under various regulatory environments without compromising their standards of data protection (Hasan et al., 2023). Making attendance management even more user-friendly, interfaces are much friendlier nowadays to make employees interact more through mobile applications and the kiosks without hitch when on the premises. Such features will help add gamified elements that encourage employees to participate and maintain a work culture that is positive—and to do this, there are incentives to gain from consistent attendance (Boutros et al., 2023). These user-centered designs make systems more accessible and also increase employee buy-in or compliance with system usage; these are very important relative to the effectiveness of the system.

The use of face recognition technology in the processes of recording attendance is gradually becoming crucial for contemporary organizations. In this technology, there is a security feature in that only an employee who is registered must mark his/her attendance, thus banning practices like buddy punching (Le et al., 2024). Face recognition systems are fast, unobtrusive and effective for processing large databases, and this latter factor particularly is a significant advantage in organizations with large numbers of staff. In addition, this way, the method can be combined with other security features, making the security system more effective and creating a solid security umbrella for organizations (Bai et al., 2023).

Further, face recognition in the attendance systems is highly efficient and hygienic, especially given the current circumstances of the world's health problems. It reduces physical contact, complies with jurisdictions' social distancing measures to protect personnel while keeping productivity levels high (Umashankar et al., 2023). In this case, the mentioned systems make use of the state-of-art algorithms to learn these changes and remain accurate and dependable. The adoption of such a modern feature not only applies innovation in the workflow, especially in the attendance management of employees, but also shows the company's concern for technological advancement as a means of raising the bar of operation safety and efficiency (Haq et al., 2023; Saraswat et al., 2023). This approach is not simply a matter of boosting the business's functioning; it is also a step in promoting the creation of a modern, safe, and effective organizational culture based on technology.

14.1.1 Problem Statement

By using traditional attendance systems, the organizations in the digital era are facing several problems. These systems consume more time that workers have to sign in physically or log in at the time of reporting, this time can be used more productively. Unintentional manual mistakes are prone to happen; they can be simple entry mistakes or intentional issues like time theft and buddy punching. Another issue is real-time verification, traditional methods are unable to predict the presence of an employee right at a moment. They usually rely on after-the-fact checking, which may affect the actual attendance. With the growth of the company, it might be difficult to manage the data amounts of manual attendance. This will be a challenge for monitoring attendance trends, accurate payroll creation and sticking to labor laws. Manual systems lack safeguards from fraud and can be easily manipulated. These issues highlight the need for an advanced, automated monitoring system which can easily deal with the modern organizational complexities and also provide accurate and immediate tracking of attendance.

14.1.2 Objectives

The main objective of having a Face Track attendance system is to apply the face recognition technology and then track attendance automatically from entry to exit which will reduce human involvement thus human errors will also be less. It increases accuracy and efficiency since it provides a quick and touchless way of marking attendance. It will provide real-time attendance which records instantly to show who is actually present. Which makes it easier to solve any discrepancies. Even if the organizational size changes it can handle it without any degradation. Biometric verification makes it impossible to falsify attendance and also supports workplace policies and regulations. It helps to understand and aggregate attendance more efficiently.

Organization: The remaining contents of this chapter are organized as follows: Section 14.2 presents some of the recent and prominent approaches on facial recognition-based attendance tracking; Section 14.3 discusses the recent innovations in facial recognition; materials used in this research is described in Section 14.4 and Section 14.5 presents the proposed approach; the results and discussion is given in Section 14.6 and the conclusions are presented in Section 14.7.

14.2 RELATED STUDIES

Ensuring an identity is another challenge in the system control especially if the communication is computer based. The other branch of biometric verification is human face recognition which has found its applications in video monitoring, human–computer interface, door control and network security among others (Rathod et al., 2025; Chempavathy et al., 2024). This paper proposes a method for a Student Attendance System that uses face recognition technology with the main algorithm of the Principal Component Analysis (PCA). This system can record the attendance of students in an environment, especially a classroom, thus minimizing errors and cases of fraud by easing the level of interaction between faculty and the student's log in and out system. Historically, the task of marking student attendance has been quite a tiresome and time-consuming exercise for college faculty. Today's biometric systems are ineffective due to the fact that there are lines to go through the fingerprint check. This paper designs an application for human attendance checking using smartphones and incorporating YOLO V3 to detect faces in the collected images and Face API from Microsoft Azure to recognize the faces. Phones with cameras take pictures at the beginning and at the close of classes to confirm attendance (Sohan et al., 2024). Students are counted and faces are recognized; a report on attendance and the number of students is presented, and monthly emails are sent to students, their parents, and faculty. This approach performs very accurately and efficiently in the implementation of real-time systems.

Attending the task of having students sign in can also be a very tiresome chore especially for the teachers. To deal with this problem, new and innovative technologically based systems of attendance taking are being adopted whereby IDs are often based on biometrics. Facial recognition which exists as one of the major biometric methods is used for multiple types of applications including video surveillance, CCTV, human–computer interfaces, indoor security access, and network security. Incorporation of facial recognition into the attendance systems, problems such as proxies and false reporting of attendance can be eliminated. Based on Eigen face values analysis, Principal Component Analysis, and Convolutional Neural Network and following the explanation of this paper, a model for auto student attendance management by facial recognition is offered (Wagh et al., 2015). The system involves identification of faces, when a face is identified, the system then analyzes it to compare with students' faces stored in the system. Indeed, such a model appears assured of achieving tremendous success in dealing with an important facet of student life, namely attendance and records. Automated systems of attendance are very vital in educational institutions particularly when the institutions use manual means where accuracy is most of the time compromised and the system of record update is tiresome. Substituting a face recognition-based attendance system is efficient in terms of the execution time when compared with other methods. The current systems use IoT and PIR sensors for attendance whereby the students use hardware devices to have their faces scanned. But the maintenance of these sensors may at times be difficult since the sensors need to be handled carefully for them not to be damaged (Ucar et al., 2024; Geetha et al., 2021). The implementation that is suggested in this paper is based on the Haar Cascade Algorithm, which proves to be highly accurate in detecting faces within a distance of 50–70 cm. Single click GUI solution enables image acquisition, dataset generation, and training of ML. If the recognition is successful, then the system provides a message regarding the student's name and roll number, while at the same time recording the time stamp in an attendance sheet. This fundamental strategy has been devised for achieving organization to the aspect of attendance in order to make it as reliable and efficient as possible when applied in education.

The moment we go through the process of roll calls in both classrooms and meetings, the main activity is utilized more by this process. Where today there is a high demand for a sophisticated, easily usable solution for the analysis of vast amounts of information that can be accessed at any time and from anywhere. This paper also presents an application called Mutabe that is specifically targeted for the use of academic faculty members (Tarmissi et al., 2024). Mutabe allows the instructors to mark everybody in their class through the capture of a single photograph of the students in their class and transforming the photo using the power of AI in face detection and recognition. Using pre-trained models, the application makes it easy to record attendance and the resulting attendance sheet is saved in an automatically created attendance database. This database is available only for the instructor using the mobile application and the website that is connected to it; therefore, simple and effective attendance tracking and management are guaranteed (Tarmissi et al., 2024).

Thus, it can be concluded that face recognition technology is important and relevant in various fields for several reasons. This paper proposes a new face recognition system that is based on the MTCNN for accurate detection, VGGFace for feature extraction, and a fast SVM for classification. Concerning real-time capability, the system performs very well in detecting and tracking multiple faces in a single frame with outstanding efficiency in attendance taking (Zhang et al., 2020). Most notably, the 'VGGFace' appears as the winner in the given benchmarks demonstrating very high accuracy, thus forming an F-score of 95% along with SVM. It amplifies that the model is efficient in appreciating facial identities; attributing the success to proper training of the model on large data sets. This research proves that the VGGFace model is very powerful particularly when worked with by several classifiers, with high accuracy being recorded from the SVM (Gu et al., 2022). Real-time face recognition-based efficient attendance systems with the help of deep learning algorithms intend to have accurate and immediate functionality in managing attendance without fail in various institutions. The critical objectives of this solution are; face detection that is very strong, proper selection of the model, efficiency in the processing, aggregation of the data, and guaranteed accuracy and reliability of the face recognition. The forms of improving the system are as follows: the usage of dlib, cvzone, and YOLO algorithms for face detection and simultaneously blur detection techniques. To sum up, this proposed system creates a possibility to solve the problem of attendance management in the conditions of a stressful working environment, which will have a high accuracy and work in real-time with the help of using effective methodologies and improving the overall value of the proposed system.

14.3 STATE-OF-THE-ART IN FACE RECOGNITION TECHNOLOGIES

Facial recognition technology has widely spread during the last few years becoming the very core of today's biometric systems. It is currently implemented in numerous applications, such as security and surveillance systems, targeted marketing and advertising, and social networks tagging. AI and ML have improved the precision, speed and flexibility of face recognition systems with great contribution. These innovations are not only experimental for upcoming technologies but they are also solving some of the basic issues with earlier technologies.

In this thesis, the state-of-the-art of face recognition as well as the new advancements and achievements of the subject matter are presented. New technologies like YOLO v9 (Wang et al., 2024) that are very useful in real-time object detection have been used to enhance the face recognition technology. Deep learning such as CNNs and GANs have elevated the accurateness and dependability of these systems to another level. Furthermore, the development of new methods of biometric recognition like pinna iris recognition and partial face recognition has enriched the sphere of identification, providing better approaches when the identification takes place in difficult conditions, for example, in the dark or when people wear masks. These advancements are supported by new approaches in multi-factor biometrics for secure facial recognition and lightweight models for edge devices making face recognition secure and available. Some of the recent technological advancements made in this field are explained in the subsequent sections.

14.3.1 YOLO V9 MODELS

The introduction of YOLOv9 (Wang et al., 2024) means that the coverage of this field is expanded in real-time object detection which in turn contributes to the improvement of the related facial recognition technologies indirectly by increasing its velocity and accuracy. Nevertheless, YOLOv9 is mainly implemented for general object detection, however, these enhancements can in fact enhance facial recognition systems significantly. The optimization termed as Programmable Gradient Information (PGI) and Generalized Efficient Layer Aggregation Network (GELAN) (Balakrishnan et al., 2024) in YOLOv9 help to get better gradient updates from the layers and to have better data paths which are very important to enhance the ability to detect as well as analyze facial characteristics under any circumstances. These improvements minimize information loss and the model evaluates more depth in the facial data, and therefore, increases the dependability and durability of the facial identification among other factors in terms of environment changes or different population. Algorithms such as YOLOv9 create the basis for numerous tech programs' improvements, including facial recognition technology, ensuring the further development of many industries' safety.

14.3.2 Pinna Iris Recognition

Iris recognition is a relatively new field of biometrics that concerns itself with the distinctive outer ear (pinna) and iris of the human eye. This two-modality approach uses the distinctiveness of both features for increasing the rate of real discrimination and security of the scans. The pinna and iris are structures that do not easily change or can almost be impossible to impersonate; that is why they are excellent biometric markers. This technology is even more useful when there is little contrast between the subject and the background or if the subject's face is partly or fully covered. For instance, in sectors of security in the pinch and the option between recognition of pinna and iris, it offers an additional security measure to make sure that only the legally allowed enter the protected areas. Besides, this method can be applicable in forensic systems for identifying people using some fragment of a biometric data set.

14.3.3 Partial Face Recognition (Face Detection with Masks)

The recent global pandemic has worsened the limitations of face recognition because many people wear masks during this period. Due to this, partial face recognition technologies have been designed to meet this challenge since they consist of eyes and forehead. These systems use sophisticated mathematical computations to check out and contrast the exterior of the face that can accurately identify the recognized individual if not a major part of the face is hidden. For instance, it is widespread to observe that most contemporary airports have applied the feature of partial face recognition in their operating security checkpoints to enhance the free and efficient movement of passengers while reducing contact with other individuals to contain the spread of infectious diseases. Such systems can also be used to sort the travelers with their corresponding identification documents regardless of the fact that the individuals wearing the masks. The creation of such technology also lies in the objectives of making the public safer as well as to keep the services running in areas where the use of masks is common.

14.3.4 3D Face Recognition

The 3D face recognition involves collection of much richer data as opposed to 2D face recognition by considering the depth of face surfaces. Unlike the previous method, this approach incorporates depth sensors and infrared cameras to make the facial contours map more precise hence eliminating the issue of lighting and facial expressions. The 3D models introduce more features than the usual 2D ones do, which help solidity and increase the accuracy of the recognition. Especially in the areas where identification is important, for example, in government facilities or banks, this technology is rather helpful. It also has application in the entertainment industry especially in the modeling of characters and animation. For instance, in gaming, 3D face recognition can increase the appearance of realism since the avatar is created in real-time and is almost an accurate mimic of the player's facial movements.

14.3.5 Multi-Factor Biometric Authentication

Multi-factor biometric authentication (MFBA) is the use of several biometric qualities to boost the security of the identification processes. This could call for combining face recognition with fingerprint, voice or iris recognition to come up with a strong course of authentication. Integrated use of multiple biometric factors greatly minimizes the chances of fraudulent access because an imposter would have to forge several distinct physiological characteristics at once. For instance, financial institutions are deploying the MFBA in order to enhance the protection of online purchases and to buy off identity theft. Since different users will be asked to provide at least two inputs of their biometrics such as a facial scan and a voice command while accessing the new application, the new interface for the banks' clients will be more secure. It also offers better security in terms of fraud protection as well as being easy to use by the intended users.

14.4 MATERIALS

This section details the materials used in this proposed approach for face recognition-based attendance management systems. The libraries and tools used in this approach are detailed here.

14.4.1 Hardware Requirements

The hardware requirements for the proposed system are primarily a camera and a computer. A high-quality camera, such as a good webcam or digital camera, capable of capturing clear video at 30 frames per second would be required for this approach. Also, requiring a fast processor (ideally with multiple cores) and at least 8GB of memory to process face recognition in real time.

14.4.2 Software and Libraries

- OpenCV: A library for image and video operations, used for face detection.
- Scikit-learn: Used for implementing and explaining the K-Nearest Neighbors (KNN) classifier.
- Flask: Provides tools and technologies to build a web server for developing a Graphical User Interface (GUI) to interact with the system.
- SQLite: A lightweight database management system used for storing user information and attendance records efficiently.
- Joblib: Used for saving and loading the KNN model.

Apart from the hardware and software requirements discussed above, an Integrated Development Environment (IDE) such as Visual Studio was used for coding and testing the application due to its powerful debugger and support for Python applications.

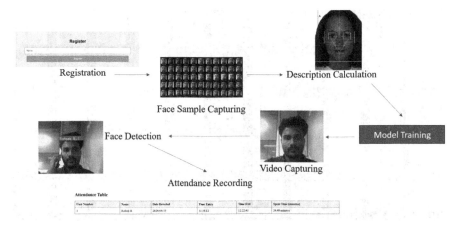

FIGURE 14.1 Overall workflow of the proposed approach.

Pip was employed for managing dependencies, ensuring all required libraries and frameworks were installable and updatable. Visual Studio facilitated the creation and management of multiple Python environments with their respective packages and versions.

This approach uses two algorithms for feature extraction and classification, such as

- Haar Cascade Classifier: In the initial phase of the application, the Haar Cascade Classifier model is used for face detection from webcam images. Proposed by Paul Viola and Michael Jones, this algorithm applies a cascade function on numerous positive and negative images for efficient object detection in real-time applications. The classifier utilizes Haar features resembling convolution kernels that highlight various facets of the face. While fast, the Haar Cascade Classifier may suffer from issues with lighting and face orientation, affecting accuracy.
- K-Nearest Neighbors (KNN): Following face detection, the K-Nearest Neighbors algorithm is employed for face recognition. KNN is a simple yet powerful classification algorithm that operates without assuming the distribution of the feature space. It calculates distances between the sample and all data examples, selects the closest k examples, and determines the majority vote of their labels. In this application, pre-processed faces are passed to the KNN model, which is trained on labeled face data to identify and label faces based on similarity in the training set.

14.5 PROPOSED APPROACH

This section discusses the proposed approach for face recognition-based attendance tracking systems. The overall workflow of the proposed approach is shown in Fig. 14.1. A detailed description of the steps involved in the proposed methodology is provided in this section.

FIGURE 14.2 A snapshot of the data collected.

14.5.1 Data Collection

Face images were collected from employees within the organization, totaling approximately 50 individuals with 100 samples each. Samples maintained veracity and diversity, with a gender distribution of 50% female and 50% male. Face detection models ensured uniformity in input data by capturing and resizing detected faces to a standard size of 50x50 pixels. A snapshot of the data collected is shown in Fig. 14.2.

14.5.2 Data Preprocessing

In the face recognition and attendance tracking application, preprocessing plays a crucial role in enhancing face detection and recognition efficiency. Initial steps include color conversion of captured images to grayscale to reduce dimensionality and normalize lighting effects. Histogram equalization improves image contrast for better detectability under varying lighting conditions. Faces are then detected using the Haar Cascade Classifier, isolated from backgrounds, cropped, and resized to a standard size for uniform input data fed into the recognition model.

14.5.3 Face Detection and Recognition Process

The system utilizes the Haar Cascade Classifier for fast and efficient face detection, removing background noise and focusing solely on facial features. Detected faces are resized and processed through the KNN algorithm for recognition, comparing features against labeled data to identify individuals accurately in real-time scenarios.

Figure 14.3 shows the process of capturing the video of a person for detecting the face and Figure 14.4 shows the face identification process.

14.5.4 Database Management and User Interface

A schema was developed in SQLite to store user information and attendance records, ensuring data consistency with SQL transactions. Timestamps for entry and exit

FIGURE 14.3 Capturing the video of a person for detecting the face.

FIGURE 14.4 The face identification process.

were recorded, calculating time spent based on the difference between entry and exit timestamps. A dynamic web interface was developed using HTML, CSS, and JavaScript in conjunction with Flask. This interface allows users to interact with the system, register new entries, and view attendance reports. Flask's lightweight nature and ease of configuration were advantageous for GUI development.

After registration, multiple images or a video of the user's face are captured to create a robust profile of their facial features. This may involve the user posing in different angles to capture a comprehensive set of facial data. Here 100 samples are collected. There is a time delay of 1 s between each capture to ensure diversity. The sample is not just collected by taking random photos. The process is backboned with the Haar cascade model to ensure faces are captured properly. The sample will only contain the face image with less background noise. In this stage the preprocessing like grayscale conversion, histogram equalization and cropping is done before saving the collected samples.

Figure 14.5 shows the database management and user interface of the proposed system.

After preprocessing the collected face image samples this major step is done. The unique facial descriptor for each registered user is computed and stored for every sample image. The face descriptions are the vector values of the positions of several distinct areas in the face. These are unique for every face. So, we have 100 sample images per person. So, the 100 descriptions are calculated. These descriptions are saved in the form of a dictionary that is key value pairs. The key (label) are the same identifiers collected during registration and the values are the list of descriptive vectors of every sample image.

Here the model used is K-Nearest Neighbors (KNN). During the training phase, the KNN model stores the feature vectors image samples (in this case, facial images) along with their corresponding labels (user identities). In the operational phase, a video capture device continuously records footage, capturing the faces of individuals when they enter the monitored area. The face detected area will be highlighted by bounding rectangles. The video should be captured in 30 fps and the quality should be maintained thoroughly. The adjustment of the camera should be done during video capturing to ensure the avoidance of light distraction and a good angle of view.

The system detects faces from the video stream in real time. This involves recognizing human faces within the video frames and possibly filtering out non-face objects. Firstly, the application uses the feature of the Haar Cascade Classifier to detect faces, which is an efficient and rather fast method that utilizes pre-trained cascades of Haar features that allow rapidly detecting the presence of faces in consecutive frames of the video stream. Once a face is detected it is removed from the background image and only the features of face enough for identification are retained. After this, the cropped face is then resized to the same scale for the sake of feeding the recognition phase the right data. For the recognition, the employed algorithm is K-Nearest Neighbors (KNN) which categorizes the detected face to the samples that exist in the training database. The KNN algorithm compares the detected face to the samples established by comparing relative features and taking the mode of the nearest neighbor samples used in labeling or identifying the face's subject. The identified name is shown on the video and the same is saved in the database also. In addition to that the detected timestamp is also recorded.

Once a face is detected, the system matches it against the stored facial descriptors to identify the individual. If a match is found, the system logs the attendance by recording the user's details along with the timestamp of entry and exit in an attendance table. Schema plan in SQLite here used to register the users and to record the attendance. A major logic is applied for getting the entry and exit timestamps. That is, from the collected timestamps of the detected object min is used for the entry timestamp in a

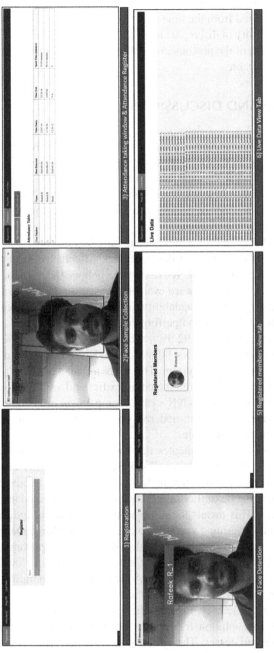

FIGURE 14.5 The database management and user interface of the proposed system.

day and max is used as the exit timestamp. The time spent is calculated from the difference of the entry and exit timestamps. The attendance table is normally composed of the user number, name, date first detected, time in, time out and total number of times taken which is obtained from the time difference between the two. This system helps eliminate the possibility of discrepancies in terms of attendance since it eliminates the manual identification of the persons in question and the recording of their presence in the format that is suitable.

14.6 RESULTS AND DISCUSSIONS

The system achieved a face detection rate of 99% and a recognition accuracy of 95%, indicating high reliability. These results underscore the system's capability to accurately detect and recognize faces under various conditions. The average recognition time was 1.9 seconds, demonstrating the system's suitability for real-time applications. This efficiency ensures minimal disruption to the users' routines. During load testing, the system maintained functionality without significant lags or critical errors, even when managing up to 100 users simultaneously. This indicates the system's scalability and robustness in handling high traffic.

To validate the effectiveness of the FaceTrack Attendance System, two machine learning models were employed: K-Nearest Neighbors (KNN) and Support Vector Machine (SVM). Both models were evaluated based on their performance metrics, including accuracy, speed, and scalability, in the context of the attendance system's requirements. The KNN model outperformed the SVM model across all key metrics. The higher accuracy (95.0% vs. 92.0%), precision (94.5% vs. 91.0%), recall (95.5% vs. 90.5%), and F1 score (95.0% vs. 90.8%) of the KNN model make it more suitable for the FaceTrack Attendance System. The KNN model's superior performance can be attributed to its ability to handle the complexities and variations in facial features among participants. Moreover, KNN's ease of implementation and real-time applicability make it an ideal choice for modern workplace environments where conditions can be dynamic and unpredictable.

The comparative analysis indicates that the KNN model is more effective than the SVM model for this application. The KNN model's higher accuracy and ease of implementation make it a better fit for the FaceTrack Attendance System. The system's robustness and flexibility suggest that it can be integrated into various organizational structures. This means it can be adapted to different industries and workplace environments, offering a versatile solution for attendance management. While the current system performs well, there is always room for improvement. Future work could focus on enhancing the system's performance in extremely low-light conditions and further reducing the recognition time. Additionally, incorporating more advanced machine learning models and exploring their potential benefits could lead to even higher accuracy and efficiency. Overall, the FaceTrack Attendance System presents a viable and efficient solution for modern organizations looking to improve their attendance tracking processes. The system's reliable performance and positive user feedback point to its potential for making a significant impact in the workplace.

FaceTrack: Real-time Attendance Marking using Face Recognition

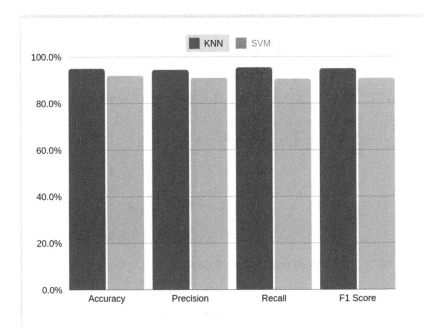

FIGURE 14.6 The precision, recall, accuracy, and f-measure comparison of KNN and SVM for the face classification.

Figure 14.6 shows the precision, recall, accuracy, and f-measure comparison of KNN and SVM for the face classification. Figure 14.7 depicts the confusion matrix for KNN classifier. The confusion matrix for the SVM classifier is shown in Figure 14.8. Figure 14.9 shows the evaluation metrics for KNN and SVM classifiers.

The KNN model dominated in all metrics across the comparison between KNN and SVM. The higher accuracy (95.0% vs. 92.0%), precision (94.5% vs. 91.0%), recall (95.5% vs. 90.5%), and F1 score (95.0 % vs. 90.8%) of the KNN model make it more suitable for the FaceTrack Attendance System. The high accuracy, precision, recall and F1 score shows that KNN can effectively mark attendance for the present participants, and it is fault tolerant in detection scenarios. The lower values for accuracy, precision, recall and F1 score of SVM as compared to KNN will lead to some false absent marking and false present marking. The performance analysis overall shows that the developed KNN model adequately performs all the required criteria for its implementation in the Face Attendance System, i.e., the FaceTrack Attendance System. Considering all the peculiarities of applying KNN and SVM, it is worth mentioning that the high accuracy of KNN and its comparative ease of implementation, as well as the fact that it can be applied in real-time, makes KNN more appropriate for complex and ever-changing conditions of the modern workplace environment.

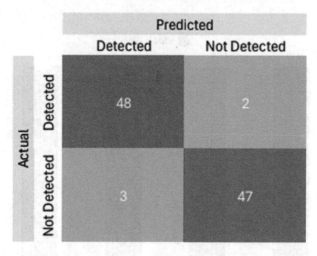

FIGURE 14.7 The confusion matrix for KNN classifier.

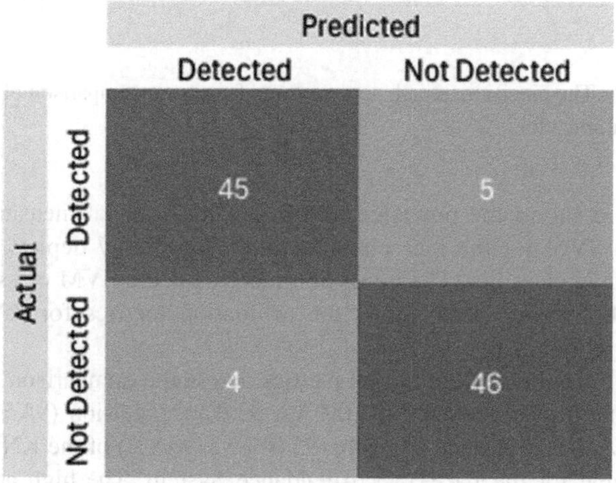

FIGURE 14.8 The confusion matrix for SVM classifier.

METRIC	KNN	SVM
ACCURACY	95.0%	92.0%
PRECISION	94.5%	91.0%
RECALL	95.5%	90.5%
F1 SCORE	95.0%	90.8%

FIGURE 14.9 The evaluation metrics for KNN and SVM classifiers.

14.7 CONCLUSIONS AND FUTURE WORK

Automated management of attendance using facial recognition technology is a well-studied problem with many potential applications in several domains. The proposed work FaceTrack Attendance System was effective in solving the limitations of traditional attendance systems which are, time consumption, less fraud tolerance, buddy punching, human error, less flexibility and more. The FaceTrack Attendance System and its progress has brought rather significant improvements in the sphere of attendance automation. The potential future research dimensions are the integration with complex systems, using hybrid approaches for advanced biometric applications, using advanced machine learning and pre-trained models for better feature extraction, and also exploring the capabilities of technologies such as blockchain for developing methods which are privacy preserving and tamper-proof.

REFERENCES

Abid, M. M., Mahmood, T., Ashraf, R., Faisal, C. N., Ahmad, H., & Niaz, A. A. (2024). Computationally intelligent real-time security surveillance system in the education sector using deep learning. *PLOS ONE*, 19(7), e0301908.

Ali, M., Diwan, A., & Kumar, D. (2024). Attendance system optimization through deep learning face recognition. *International Journal of Computing and Digital Systems*, 15(1), 1527–1540.

Bai, J., Zhang, X., Song, X., Shao, H., Wang, Q., Cui, S., & Russello, G. (2023, October). CryptoMask: Privacy-preserving face recognition. In *International Conference on Information and Communications Security* (pp. 333–350). Springer Nature Singapore.

Balakrishnan, T., & Sengar, S. S. (2024). RepVGG-GELAN: Enhanced GELAN with VGG-STYLE ConvNets for Brain Tumour Detection. arXiv preprint arXiv:2405.03541.

Boutros, F., Struc, V., Fierrez, J., & Damer, N. (2023). Synthetic data for face recognition: Current state and future prospects. *Image and Vision Computing*, 135, 104688.

Budiman, A., Yaputera, R. A., Achmad, S., & Kurniawan, A. (2023). Student attendance with face recognition (LBPH or CNN): Systematic literature review. *Procedia Computer Science*, 216, 31–38.

Chempavathy, B., Dhanalakshmi, M., Varun, M. G., Bohra, V., & Kalsariya, V. A. (2024, March). An improved attendance monitoring system through facial recognition using RetinaFace algorithm. In *2024 3rd International Conference for Innovation in Technology (INOCON)* (pp. 1–5). IEEE.

Chen, L. (2024). Analysis of facial recognition attendance technology based on artificial intelligence algorithms in political course e-learning teaching. *Entertainment Computing*, 52, 100821.

Chen, X. (2023, May). Study on student attendance system based on face recognition. In *Journal of Physics: Conference Series* (Vol. 2492, No. 1, p. 012015). IOP Publishing.

Dang, T. V. (2023). Smart attendance system based on improved facial recognition. *Journal of Robotics and Control (JRC)*, 4(1), 46–53.

Feroze, S. A., & Ali, S. Z. (2024). The facial recognition technology in academic attendance: A comparative study for real-time management. *International Journal of Technology, Innovation and Management (IJTIM)*, 4(1), 1–19.

Geetha, V., Anbumani, V., Selvi, T., Sindhuja, C. S., & Vanathi, S. (2021, February). IoT based well-organized hostel power consumption and attendance administration system. In *IOP Conference Series: Materials Science and Engineering* (Vol. 1055, No. 1, p. 012015). IOP Publishing.

Gu, M., Liu, X., & Feng, J. (2022). Classroom face detection algorithm based on improved MTCNN. *Signal, Image and Video Processing*, 16(5), 1355–1362.

Gururaj, H. L., Soundarya, B. C., Priya, S., Shreyas, J., & Flammini, F. (2024). A comprehensive review of face recognition techniques, trends and challenges. *IEEE Access*.

Hasan, M. R., Guest, R., & Deravi, F. (2023). Presentation-level privacy protection techniques for automated face recognition—A survey. *ACM Computing Surveys*, 55(13s), 1–27.

Haq, H. B. U., & Saqlain, M. (2023). Iris detection for attendance monitoring in educational institutes amidst a pandemic: A machine learning approach. *Journal of Industrial Intelligence*, 1(3), 136–147.

Irawati, A. R., Kurniawan, D., Utami, Y. T., & Taufik, R. (2024). An exploration of TensorFlow-Enabled convolutional neural network model development for facial recognition: Advancements in student attendance system. *Scientific Journal of Informatics*, 11(2), 413–428.

Jha, P. B., Basnet, A., Pokhrel, B., Pokhrel, B., Thakur, G. K., & Chhetri, S. (2023). An automated attendance system using facial detection and recognition technology. *Apex Journal of Business and Management*, 1(1), 103–120.

Kamil, M. H. M., Zaini, N., Mazalan, L., & Ahamad, A. H. (2023). Online attendance system based on facial recognition with face mask detection. *Multimedia Tools and Applications*, 82(22), 34437–34457.

Kumar, P., TA, S. L., & Santhosh, R. (2023, May). Face recognition attendance system using local binary pattern algorithm. In *2023 2nd International Conference on Vision Towards Emerging Trends in Communication and Networking Technologies (ViTECoN)* (pp. 1–6). IEEE.

Le, M. H., & Carlsson, N. (2023, April). IdDecoder: A face embedding inversion tool and its privacy and security implications on facial recognition systems. In *Proceedings of the Thirteenth ACM Conference on Data and Application Security and Privacy* (pp. 15–26). ACM.

Nguyen-Tat, B. T., Bui, M. Q., & Ngo, V. M. (2024). Automating attendance management in human resources: A design science approach using computer vision and facial recognition. *International Journal of Information Management Data Insights*, 4(2), 100253.

Opanasenko, V. M., Fazilov, S. K., Mirzaev, O. N., & Sa'dullo ugli Kakharov, S. (2024). An ensemble approach to face recognition in access control systems. *Journal of Mobile Multimedia*, 749–768.

Rathod, V. M., & Agarwal, S. (2025). A study on performance enhancement of deep learning-based face detection system. In *Artificial Intelligence and Information Technologies* (pp. 524–529). CRC Press.

Rathore, K. S., Pandey, A., Gupta, A., Srivastava, D., Agrawal, K., & Srivastava, S. (2024, January). Design and implementation of efficient automatic attendance record system based on facial recognition technique. In *AIP Conference Proceedings* (Vol. 2978, No. 1). AIP Publishing.

Saraswat, D., Bhattacharya, P., Shah, T., Satani, R., & Tanwar, S. (2023). Anti-spoofing-enabled contactless attendance monitoring system in the COVID-19 pandemic. *Procedia Computer Science*, 218, 1506–1515.

Shukla, A. K., Shukla, A., & Singh, R. (2024). Automatic attendance system based on CNN–LSTM and face recognition. *International Journal of Information Technology*, 16(3), 1293–1301.

Sohan, M., Sai Ram, T., Reddy, R., & Venkata, C. (2024). A review on yolov8 and its advancements. In *International Conference on Data Intelligence and Cognitive Informatics* (pp. 529–545). Springer, Singapore.

Tarmissi, K., Allaaboun, H., Abouellil, O., Alharbi, S., & Soqati, M. (2024, January). Automated attendance taking system using face recognition. In *2024 21st Learning and Technology Conference (L&T)* (pp. 19–24). IEEE.

Ucar, C., Maloku, M., Yugay, O., & Budimir, D. (2024, June). IoT motion detection sensors for monitoring in a smart campus. In *2024 13th Mediterranean Conference on Embedded Computing (MECO)* (pp. 1–4). IEEE.

Umashankar, B. S., Lakshmi Narayan, S., Ruthvik, M., & Deshpande, P. (2023, February). Facial recognition-based attendance and smart COVID-19 norms monitor. In *ICDSMLA 2021: Proceedings of the 3rd International Conference on Data Science, Machine Learning and Applications* (pp. 571–579). Springer Nature Singapore.

Wagh, P., Thakare, R., Chaudhari, J., & Patil, S. (2015, October). Attendance system based on face recognition using eigen face and PCA algorithms. In *2015 International Conference on Green Computing and Internet of Things (ICGCIoT)* (pp. 303–308). IEEE.

Wang, C. Y., Yeh, I. H., & Liao, H. Y. M. (2024). Yolov9: Learning what you want to learn using programmable gradient information. arXiv preprint arXiv:2402.13616.

Zhang, N., Luo, J., & Gao, W. (2020, September). Research on face detection technology based on MTCNN. In *2020 International Conference on Computer Network, Electronic and Automation (ICCNEA)* (pp. 154–158). IEEE.

15 Optimizing Information Systems for Green Computing in Higher Education
From Awareness to Action

Tanvir Chowdhury, Habibur Rahman,
Monjurul Islam Sumon, and Md Fokrul Akon

15.1 INTRODUCTION

The utilization of information and communication technologies (ICTs) in higher education institutions has experienced a significant and rapid increase in recent years. ICTs are extensively included throughout all aspects of higher education, ranging from online learning platforms to administrative software systems. Nevertheless, this growing digitalization is accompanied by substantial environmental consequences.

Electronic gadgets and data centers utilized in educational environments consume substantial quantities of electricity, hence contributing to the release of greenhouse gases. Moreover, the production, utilization, and elimination of ICT equipment result in significant amounts of electronic waste. Consequently, it is imperative for higher education institutions aiming to function in an environmentally sustainable manner to tackle the ecological impact of computing.

Green computing encompasses the practice of utilizing computers and other resources in a way that preserves the environment. Eco-friendly computing encompasses the process of designing, manufacturing, utilizing, and disposing of computing systems in a manner that minimizes their environmental impact. Several fundamental ideas of green computing encompass:

- Energy efficiency – Achieving energy savings by utilizing hardware elements and controlling power settings that are designed to consume less energy.
- Proper dumping and recycling – Ensuring the responsible disposal and recycling of electronic waste.
- Green data centers are designed to achieve optimal energy efficiency by implementing strategies such as efficient cooling systems, server virtualization, and the utilization of renewable energy sources.

- Minimizing paper waste – Implementing printer settings and workplace procedures to reduce paper printing.
- Sustainable manufacturing – Utilizing ecologically responsible techniques for producing and delivering computer components

Implementing environmentally friendly computing standards is crucial for reducing the substantial carbon emissions generated by the information technology sector.

Importance of understanding green computing for IT professionals and students

Comprehending green computing principles is crucial for IT professionals, both present and future, due to the significant and expanding environmental influence of the information technology sector. The IT sector accounts for around 2% of worldwide CO2 emissions, which is comparable to the emissions produced by the aviation industry. E-waste is the most rapidly expanding waste stream globally.

IT professionals, as producers of computer hardware and software, bear the obligation of designing energy-efficient, durable, and recyclable products. Having a deep understanding of sustainability enables individuals to make ecologically conscious choices when it comes to product creation, development, and infrastructure management.

Students embody the upcoming cohort of IT experts who have the potential to lead the development of innovative environmentally friendly technology and services. By introducing sustainability at an early stage, individuals are able to integrate environmentally conscious thoughts into their studies and future professional endeavors.

In general, having a solid foundation in green computing enables both existing and aspiring IT professionals to effectively address urgent environmental problems through their professional endeavors.

15.2 LITERATURE REVIEW

15.2.1 Discuss the Increasing Use of ICTs in Higher Education Institutions and the Potential Environmental Impact

Higher education institutions today rely heavily on ICTs like computers, servers, networking equipment, and data centers. A 2021 study found that on average, universities spend over $300 per student annually just on software licenses and cloud computing services [1]. This increasing digitalization enables innovative teaching and research methods but also results in significant energy use and electronic waste.

According to one estimate, the ICT ecosystem accounts for 2–5% of global greenhouse gas emissions – a share that could grow to 14% by 2040 [2]. Much of this impact comes from the vast amounts of electricity needed to power data centers and computing infrastructure. Reducing the environmental footprint of ICTs is therefore crucial for limiting global warming.

15.2.2 Highlight the Importance of Addressing Green Computing Awareness and Adoption in Educational Settings

Research shows that while students and staff in higher education generally have a high level of environmental awareness, adoption of green computing practices remains low [3]. This highlights the need to bridge the gap between awareness and action by promoting green computing policies and initiatives tailored to educational institutions. As centers of research, innovation, and learning, universities are uniquely positioned to lead by example in sustainable technology use. Implementing green computing in areas like energy-efficient data centers, responsible e-waste disposal programs, and environmentally preferable purchasing policies can significantly reduce the sector's IT footprint while educating students on sustainability.

By addressing green computing holistically across campus operations, administration, academics, and research, higher education institutions can drive positive environmental change and produce sustainability-focused graduates ready to lead in the 21st century.

Green computing is the utilization of computing resources in an environmentally sustainable manner while ensuring optimal computing performance. Global warming refers to the long-term increase in Earth's average surface temperature due to human activities, such as the burning of fossil fuels and deforestation. The Earth's climate system is experiencing a persistent increase in average temperature, which is attributed to various factors. Scientific comprehension of the multiple factors contributing to global warming has been steadily advancing over the past decade. Climate change and its related effects exhibit geographical variability worldwide. Currently, weather patterns are becoming erratic worldwide. The United Nations Framework Convention on Climate Change (UNFCCC) is diligently striving to fulfill its goal of averting hazardous anthropogenic (human-caused) climate change. Due to the phenomenon of global warming, makers of IT equipment are compelled to comply with a range of energy needs as stipulated by regulations and legislation pertaining to environmental norms. Green computing is an environmentally conscious and sustainable approach that aims to create a healthier and safer environment while still meeting the technology requirements of present and future generations. This study provides a comprehensive overview of significant literature pertaining to the topic of green computing, highlighting the crucial role of green computing in promoting sustainable development [4].

The advent of information and communication technology (ICT) has revolutionized all aspects of our lives, including employment, education, and leisure. However, it is important to acknowledge that ICT also has significant environmental implications. The increasing demand for computer literacy in both public and private sectors has led to numerous employment opportunities worldwide. The computer's capacity to efficiently store, retrieve, and manipulate vast quantities of data at a low cost has resulted in its extensive utilization for managing various clerical, financial, and service documentation tasks within enterprises. However, throughout every phase of a computer's lifespan, starting from its manufacturing, continuing through its usage, and concluding with its disposal, it presents many environmental issues.

Multiple scientists and authors have cited their findings on information and communication technology (ICT) and its influence on the environment. However, the question of whether green computing is truly successful in promoting environmentally friendly and sustainable IT is still unresolved. This article discusses the awareness of green computing and provides an overview of the main areas where IT businesses can save energy and reduce costs. Furthermore, we explore a systematic methodology of green computing, including its standardization and adherence to regulations, as well as the obstacles it faces [5]. The IT systems are designed to identify patterns in existing algorithms and datasets and provide appropriate solution concepts [6]. IDS functions as a real-time surveillance system capable of detecting suspicious activity and issuing alerts in the event of unauthorized access or malicious assaults [7]. The primary objective of this project is to use VGG19 and CNNs to classify sensor photos [8].

15.3 METHODOLOGY

Strategies for Implementing Green Computing in Higher Education
Introducing green computing in higher education institutions necessitates a thorough and sustainability-oriented approach. An essential element involves formulating a green computing strategy that advocates for environmentally conscious laws and practices.

15.3.1 CREATE A SUSTAINABLE GREEN COMPUTING PLAN

A sustainable green computing plan establishes a structure for incorporating environmentally friendly initiatives into operations. The plan should delineate objectives to diminish energy consumption, appropriately discard electronic trash, procure environmentally friendly technologies, and enlighten the college community [9].

For example, the plan could aim to cut data center energy use by 15% in 3 years through server virtualization, shutdown policies, and HVAC optimizations. It is critical to get stakeholder buy-in across IT, facilities, procurement, and administration when creating the plan.

15.3.2 ENSURE PROPER E-WASTE DISPOSAL

Colleges and universities must establish e-waste disposal procedures aligned with environmental regulations. This involves providing easily accessible recycling stations and coordinating pickup of discarded electronics.

Special attention should be paid to hazardous components in e-waste like batteries, lamps, and cathode ray tubes. Staff training is essential to sort materials correctly for safe transportation and processing by qualified recycling partners.

15.3.3 Make Eco-Conscious IT Purchasing Decisions

Purchasing decisions should account for sustainability by evaluating energy efficiency, materials, and end-of-life considerations. Priority can be given to suppliers with green credentials like EPEAT registration and ENERGY STAR certification.

For example, desktops, monitors, and printers meeting EPEAT Silver criteria can reduce greenhouse gas emissions by 30% or more over less eco-friendly models. Making sustainable IT purchases demonstrates a commitment to green computing.

15.4 RESULT ANALYSIS

Paths to Address Environmental Effects of Computing in Higher Education

As computing technology continues to advance and become more ubiquitous in higher education institutions, it is crucial that we examine the environmental impacts associated with it. Murugesan has proposed four key paths that can help address these effects: green use, green disposal, green design, and green manufacturing.

Green Use of Computing Technology

Educational institutions can adopt policies and execute best practices that prioritize the sustainable and environmentally friendly utilization of computers and other information and communication technology (ICTs). This includes encouraging practices like:

- Using energy-saving settings on devices
- Minimizing paper printing
- Powering down equipment when not in use
- Transitioning storage and software systems to cloud-based services to reduce local hardware needs
- Promoting awareness of green use among students and staff is also important

Green Disposal of Electronic Waste

Proper disposal of broken or outdated ICT equipment is key. Educational institutions should have clear policies and programs in place focused on:

1. Reusing still functional older equipment where possible through donations etc.
2. Recycling e-waste in an environmentally responsible manner
3. Disposing of non-recyclable e-waste properly to avoid toxicity issues

Making it convenient for staff and students to follow green disposal practices is essential.

Integrating Green Computing Paths into Higher Education

In addition to considering the environmental impact of using and disposing of resources, educational institutions should also assess their procurement strategies, equipment design choices, and collaborations with manufacturers. To mitigate environmental impacts associated with ICTs, higher education institutions can implement complete green computing programs by evaluating all four pathways suggested by Murugesan. Acquiring support and involvement from various stakeholders both within and outside of educational institutions will be necessary for this. However, the enduring advantages for sustainability justify the endeavor's value.

The Role of Educational Institutions, IT Community, and Legislative Bodies

It is imperative for educational institutions to assume responsibility for spearheading the adoption of ecologically sustainable computer activities. Colleges and universities, as educational institutions that prepare students for the future, should demonstrate environmental accountability by implementing energy-efficient IT infrastructure and establishing appropriate protocols for disposing of electronic waste. These activities function as a model and have extensive consequences across society.

Promoting Eco-Friendly ICT Usage and Disposal

- Transition to ENERGY STAR-certified computers, displays, printers and other office equipment
- Establish campus-wide policies for responsible disposal of e-waste
- Educate students and staff about green computing best practices
- Develop IT sustainability plans aligned with campus sustainability goals

The IT community also plays a crucial role. Technology suppliers and service providers should persist in their efforts to provide new and improved solutions that enhance efficiency and environmental sustainability. They have the option to collaborate with educational institutions in order to execute customized solutions that align with their green computing goals. Legislative bodies possess the chance to promote and support sustainability. Policymakers have the ability to allocate financial resources, offer rewards, and establish directives to support schools in implementing more environmentally friendly technologies. Effective policies facilitate the elimination of obstacles to adopting environmentally friendly practices.

Potential Impact

Implementing green computing practices in education can:

- Reduce energy usage, costs and greenhouse gas emissions. Minimize e-waste generation through reuse and recycling. Set an example for students that influences lifestyle choices. Demonstrate institutional values of environmental stewardship. The collective impact of many schools adopting greener computing is immense.

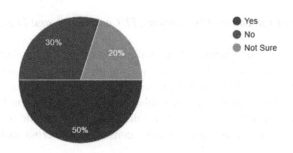

FIGURE 15.1 Awareness of Green computing initiatives.

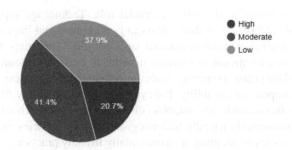

FIGURE 15.2 Understanding of Green computing concepts.

Figure 15.1 represents a pie chart illustrating the outcomes of a survey conducted to assess the level of awareness regarding green computing activities among university students. Among the group of 30 participants: 50% (15 individuals) possess knowledge about green computing initiatives.9 individuals, constituting 30% of the total, lack awareness.6 individuals, constituting 20% of the total, express uncertainty over the existence of any efforts or programs of this nature. The chart facilitates the visualization of the proportions pertaining to the respondents' levels of awareness.

The pie chart in Figure 15.2 depicts the self-assessed comprehension of green computing principles among a group of 29 participants. The allocation is as follows: 41.4% of respondents consider their level of understanding to be high. 37.9% consider it to

Opinions on University's Efforts in Green Computing: How would you rate the university's efforts in promoting green computing practices?

30 responses

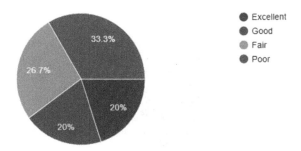

FIGURE 15.3 Opinion on university's efforts in Green computing.

Encouragement for Green Computing Research: Do you think the university provides enough encouragement and support for research focused on advancing green computing technologies and practices?

30 responses

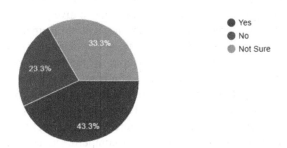

FIGURE 15.4 Encouragement for Green computing Research.

be moderate. 20.7% of individuals assess their level of comprehension as being low. The chart illustrates the self-assessed knowledge of the respondents on green computing, indicating that most of them perceive their understanding of the topic to be moderate to high.

The pie chart in Figure 15.3 illustrates the survey responses of 30 persons regarding their perspectives on their university's initiatives in encouraging environmentally friendly computing habits. 33.3% of respondents rated the efforts as Excellent.

26.7% consider them to be of good quality. 20% of respondents consider them to be Fair. 20% of respondents rated them as Poor. According to this figure, most of the participants have a positive perception of the university's efforts, since 60% of them rated them as either excellent or good. Nevertheless, a substantial proportion (40%) holds the belief that the endeavors are merely satisfactory or inadequate.

Importance of Green Computing in University Decision-Making: How important do you believe green computing considerations are in the university's decision-making processes, particularly in technology-related matters?

30 responses

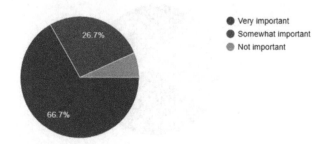

FIGURE 15.5 importance of green computing in university decision.

The pie chart in Figure 15.4 assesses perspectives of the university's provision of sufficient motivation and assistance for research in green computing. 43.3% of participants hold the belief that the university does offer sufficient support. A total of 33.3% of respondents believe that the university does not offer adequate help. 23.3% of individuals express uncertainty about the extent of encouragement and funding for research on green computing. This indicates a divergence in viewpoints, with a majority expressing agreement and a substantial portion either disagreeing or unsure.

The pie chart in Figure 15.5 depicts the viewpoints of 30 participants regarding the significance of green computing factors in their university's decision-making procedures, particularly in relation to technology. A significant majority of 66.7% believe green computing factors to be of utmost importance. 26.7% consider them to be moderately significant. Approximately 6.6% of individuals, as estimated from the graph due to the absence of the actual figure, hold the belief that they are not significant. A substantial majority prioritize green computing as a crucial determinant in the university's technology-related choices.

REFERENCES

[1] Campus Computing,"Cloud Computing in Higher Education Report," 2021, www.campuscomputing.net.
[2] L. Belkhir and A. Elmeligi, "Assessing ICT Global Emissions Footprint: Trends to 2040 & Recommendations," *Journal of Cleaner Production*, Volume 177, 2018, pp. 448–463.
[3] S. Murugesan, *Promoting Sustainable Green Computing and Green IT Practices*. IT Professional Magazine, 2015.
[4] B. Saha, "Green Computing," *International Journal of Computer Trends and Technology (IJCTT)*, Volume 14, Number 2, August 2014, ISSN: 2231-2803, p. 46, www.ijcttjournal.org.

[5] K. Raza, V. K. Patle and S. Arya, "A Review on Green Computing for Eco-Friendly and Sustainable IT," *Journal of Computational Intelligence and Electronic Systems*, Volume 1, Number 1, June 2012, pp. 3–16.

[6] M. T. Chowdhury et al., "Comparison of Feasibility between Machine Learning Algorithms In Terms of Predicting Energy Consumption of Smart Grid," *2023 4th International Conference on Big Data Analytics and Practices (IBDAP)*, Bangkok, Thailand, 2023, pp. 1–6, doi: 10.1109/IBDAP58581.2023.10272003.

[7] M. S. Hossain et al., "Performance Evaluation of Intrusion Detection System Using Machine Learning and Deep Learning Algorithms," *2023 4th International Conference on Big Data Analytics and Practices (IBDAP)*, Bangkok, Thailand, 2023, pp. 1–6, doi: 10.1109/IBDAP58581.2023.10271964.

[8] M. T. Chowdhury, H. Rahman, M. I. Sumon and A. Talha, "Classification of satellite images with VGG19 and Convolutional Neural Network (CNN)," *2024 2nd International Conference on Advancement in Computation & Computer Technologies (InCACCT)*, Gharuan, India, 2024, pp. 397–402, doi: 10.1109/InCACCT61598.2024.10551185.

[9] I. S. Jacobs and C. P. Bean, "Fine Particles, Thin Films and Exchange Anisotropy," *Magnetism*, vol. III, G. T. Rado and H. Suhl, Eds. New York: Academic, 1963, pp. 271–350.

16 Evaluating Thin Client Solutions

A Sustainable and Energy-Efficient Alternative to Traditional Classroom PCs

Md Tanvir Chowdhury, Habibur Rahman, Monjurul Islam Sumon, Md Mahmud Hasan Akhnoda, and Samia Shameem Haque

16.1 INTRODUCTION

In today's information-driven era, the integration of modern technology in the education sector has become a vital necessity. The developed and developing countries have adopted all the advancements of these technologies in their education sectors. To boost educational levels in developing countries, an affordable and long-lasting computing environment must be implemented. Like other developing countries, almost every educational institution in Bangladesh uses computers for all purposes of work. From handling all the procedures of admission to taking attendance in class computers are being used. Using these technologies has become essential. However, traditional PCs' high energy consumption and cost create challenges. Consumption of higher energy of these traditional PCs causes great harm to the environment. The most significant environmental impact of PC energy use is the atmospheric release of greenhouse gases like carbon dioxide (CO_2).

Since not all of these tasks require a high-end computer, some of them can be performed with a basic computer that uses less energy. Therefore, a replacement of these higher-energy-consuming devices can create a positive impact on the environment. Raspberry Pi can be a great alternative to the traditional PCs.

Raspberry Pi is a multi-purpose low-cost Advanced Reduced-Instruction-Set-Computer (ARM) processor-based miniature device that has been utilized as a standalone machine in schools to improve the education provision in rural areas [1]. Raspberry Pi is a low-cost, credit-card-sized single-board computer. Similar to other types of computers, RPi consumes power during its operation. This includes the power needed to operate the hardware components and to perform tasks or run software on the platform [2]. It relatively consumes lower average power as compared to the desktop computer and laptop and has the potential to save both money and energy due to its low cost and relatively low power consumption.

For the completion of some tasks that can be done with those devices that can have less energy consumption, Raspberry Pi can be a great replacement for it. This study suggests using Raspberry Pi to reduce energy consumption and to take steps toward the advancement of green computing. The smart grid (SG) refers to the integration of sensors and software into an existing electrical infrastructure, which enables utilities and individuals to access extra information and react more quickly to changes in energy [3].

Also, machine learning and deep learning are specialized branches of computer science (AI) that are particularly effective at processing large amounts of data [4]. The primary objective of this project is to use VGG19 and CNNs to classify sensor photos [5].

16.1.1 Background

In recent years, there has been a rise in technological developments targeted at improving student learning experiences. Traditional classroom PCs are now considered standard in educational institutions as a result of the rising need for computer-based learning resources. But the high consumption of energy in the usage of these resources has a significant impact on the environment.

The electricity required to power traditional PCs is often generated by burning fossil fuels, such as coal, oil, and natural gas. The combustion of these fuels releases greenhouse gases, particularly carbon dioxide (CO_2), into the atmosphere. These emissions are a major contributor to global climate change and the resulting impacts, such as rising temperatures, extreme weather events, and sea-level rise. Along with huge amounts of resources like metals, minerals are also needed which are harmful to the environment.

This study is focused on minimizing the consumption of energy from classrooms where high-power-consuming devices are used for doing minimalistic tasks by replacing them with low-power-consuming devices such as Raspberry Pi.

16.2 LITERATURE REVIEW

In a study, a comparative analysis has been done between Raspberry Pi (RPi) and other devices for the calculation of energy consumption. The experiment has been done in these devices for the same 20 operations. The outcome of the study shows that the RPi consumed relatively lower average power as compared to the desktop computer and laptop [2]. Another study [2] offers a multi-layer approach to increase the energy efficiency of thin client systems. By dynamically adjusting radio sleep times and operating at many protocol levels, it effectively decreases energy consumption while maintaining the experience of users. This improvement in thin client energy-efficient computing is notable since it has undergone thorough experimental validation demonstrating its practicality. This research sets a positive example for holistically tackling energy issues by demonstrating how cross-layer collaboration could enhance power-conscious computing in thin client scenarios.

This study [6] investigates the development of a Raspberry Pi-based thin client system that is tailored for educational settings. The study makes use of the flexibility and affordability of the Raspberry Pi to meet the critical need for affordable computing solutions in educational settings. The study adds to the corpus of knowledge on thin client applications in education by examining the system's design and evaluating its usability. This innovative use of Raspberry Pi emphasizes its potential to close the digital divide in educational institutions by providing students with access to computing resources while maintaining cost and energy efficiency.

Another study [7] explores how thin-client computing and energy efficiency are related. It seeks ways to use less energy while still performing at its best. Analysis of the energy implications of thin clients gives critical new information on green computing practices. In order to provide the groundwork for future developments in green computing and the reduction of environmental impact, the research emphasizes the requirement of green designs in the environment of remote computing systems. This study [8] investigates the energy efficiency of the computer cluster, an essential component of the computing infrastructure. the analysis of performance enhancements, power consumption patterns, and how these affect energy use. The Raspberry Pi is a fantastic example of how adopting energy-efficient components can assist sustainable computing. The findings underscore the importance of incorporating energy-conscious methods in clustered computing systems, giving businesses guidance on how to boost computational efficiency while minimizing their environmental impact.

This study uses the Raspberry Pi 2 Model B to investigate the potential of low-cost computing. In an effort to inform readers about the potential uses of this low-cost computing solution in a variety of industries, it investigates the capabilities and performance of the system. The system's processing capacity, usability, and cost-effectiveness are all carefully examined in the study. By demonstrating the Raspberry Pi 2 Model B's capacity to carry out a variety of computer tasks while adhering to a strict budget, this article adds to the larger conversation on affordable computing solutions.

16.3 METHODOLOGY

16.3.1 Data Collection Methods

For this study, both quantitative and qualitative analyses were required. A detailed data collection method has been used to thoroughly explore the power consumption of classroom PCs and Raspberry Pi devices as well as to analyze educators' perspectives on the adoption of thin clients as a workable replacement for traditional PCs in the classroom. Several variables are required to measure power usage for quantitative data. Such as type of computer, CPU, RAM, HDD, monitor, usage duration, etc. For this study, both qualitative and quantitative data were collected from East West University.

The power consumption mainly occurs in a computer from motherboard, RAM, memory, CPU, GPU. We gathered the information of these units from the computers provided in the classrooms of East West University. We needed to calculate the total number of classrooms on the campus and the class schedule for the duration of the total class period held in each classroom as well. For the number of total classrooms, only regular classrooms were considered excluding the labs and seminar rooms. Based on the investigated values, the total power consumption of those classroom PCs was calculated along with the comparison of the power consumption of the Raspberry Pi system if the traditional computers were replaced in the classrooms. For the qualitative analysis, we use a representative sampling model [9]. Interviews were taken from East West University. The interviews were taken from the faculties from all departments who take classes in the regular classrooms of the campus. The motive of the interview sessions was to get an idea of the daily usage of the classroom PCs used by the faculties and their opinion about the replacement of traditional PCs with Raspberry Pi in the classroom. The interview questions were mainly prepared in such a way that they achieve objective 2 of this study.

Interview questions that were asked:

1. Do you use a classroom PC in every class you take?
2. Do you get all the applications provided in the PC for taking the class?
3. Do you have any idea of a thin client (Raspberry Pi)?
4. Have you ever used Raspberry Pi?
5. If the Authority decides to replace the classroom PC with the thin client would it be comfortable to use it?
6. Can you share your opinion about the replacement of traditional PCs with Raspberry Pi in the classroom?

16.3.2 Participants

Various studies have collected data on power consumption on a large scale. A study [10] shows the energy usage for miscellaneous electric loads in offices and commercial buildings. The data collection scale of this study was limited to East West University.

For the data collection of the quantitative information, the classrooms of East West University were visited. The information was collected physically in front of the classroom messengers with permission. The total number of classrooms was collected from the messengers on each floor of the university campus.

For the quantitative information, the interviews were taken from the faculties from all departments of the university. faculties were delighted to express their opinion regarding the study.

16.3.3 Data Analysis

Our data analysis is categorized into two parts. Firstly, quantitative analysis was made based on the usage of computers in regular classrooms. Considering some usual activity tasks done with desktop computers during the period of time of classes and the duration of using the computers the analysis is done. Regarding this, data of the time period and total power consumption while using these tasks were noted down. And the total amount of power consumption was calculated with the total number of classrooms. Furthermore, Raspberry Pi 4 Model B (Broadcom BCM2711) devices were needed and tested by loading similar activity tasks to compare the power consumption by calculating with the same procedures for the classroom's desktops. Secondly, quantitative data analysis was performed from the interview sessions with the faculties. Interviews were held with all the departments' faculties as the classrooms of East West University are considered to be "multi-Purpose Classroom", all the faculties use similar classrooms. However, almost all the facilities of the university come to make use of the same environment in the classroom while taking their classes. The usage of the classroom PCs of the faculties while taking classes, types of applications and software are used by faculties while taking the classes and their opinion regarding the changes in the environment of classrooms with the replacement of PCs with Raspberry Pi were recorded.

16.3.4 Research Ethics

The data collection provided by the university messengers was confirmed personally through physical inspection along with permission. Thus there is no scope for error in information.

Furthermore, interviewing ethics were followed throughout the data collection process. All participants were given a consent document detailing the study and the purpose of the interview. The interview was only conducted if all of the terms and conditions were agreed upon and signed the consent form.

16.4 EXPERIMENTAL RESULT AND DISCUSSION ANALYSIS

The data relating to the consumption of energy by classroom PCs were collected through physical visits to the classrooms of East West University. The specifications of the classroom computers were collected along with the total number of classrooms where only regular classes are held along with the duration of total classes in each classroom. Following an in-depth analysis of these configurations, it was found that all classrooms use the same sorts of computers for at most 8 hours every day, from 8:30 a.m. to 4:40 p.m.

Visiting the campus, a total of 65 regular classrooms were found with 9 classrooms on the ground floor, 12 classrooms on the first floor, 13 classrooms on the third floor, 13 classrooms on the third floor and 24 classes in FUB (FarashUddin Building) of East West University. These classrooms have computers with the same configuration and almost all normal classes are held in all departments where only class lectures are delivered. Therefore, the desktops that are used in the classrooms are replaceable. We have shown the data that we have collected from the classroom in Table 16.1.

Analysis shows that the computers in the classroom are used for at least 8 hours. So the energy consumption for a total of 65 classrooms would be around:86×8×65 Wh=44.72 KWh as per the calculation. For the comparison, we used Raspberry Pi 4 Model B to measure the power consumption. The specification of the Raspberry Pi 4 Model B for the comparison is given in Table 16.2.

Sustainable Thin Client Solutions for Energy-Efficient Classrooms

TABLE 16.1
Total power consumption of each classroom's PC

Processor	HDD	RAM	Monitor	Power Consumption
Intel core™ i7-10700	Toshiba 1TB	8 GB	HP P22va G4 21.5 inch Full HD Monitor	86 W

TABLE 16.2
Raspberry Pi 4 Model B specifications

Processor	Clock rate	RAM	Monitor
Broadcom BCM2711	1.5GHz	4 GB	HP M22f Full HD 21.5 Inch Monitor

TABLE 16.3
Power usage of Raspberry Pi module in various states

Idle(A)	Software in use state (A)	App Standby state (A)	Volt (V)
0.26	0.61	0.28	5.23
0.25	0.63	0.29	5.23
0.27	0.61	0.31	5.25
0.30	0.65	0.31	5.28
0.31	0.68	0.33	5.29
0.26	0.65	0.29	5.32
0.26	0.63	0.30	5.29
0.27	0.63	0.30	5.25
Avg=0.27	Avg=0.27	Avg=0.27	Avg=0.27

To measure the power consumption the KEWEISI Voltage Current Detector was used. This detector displayed the ampere consumed by the Pi module each second. We collected snapshots of the consumption in each state and stage. We ran the Microsoft Office 365 web app to imitate the stress on the device and then measured the power consumption in real time (Table 16.3).

Raspberry Pi consumed around 1.42 W when it was running in idle mode while no app was running in the background. It consumes 3.37 W when the MS Office web app was running, and in app standby mode when the app was running in the background it consumes 1.58 W. We observe that, even if the Raspberry Pi is in a working state all the time, the power consumption is far less than traditional PCs. We added the average power consumption of the monitor 20 W with the power consumption

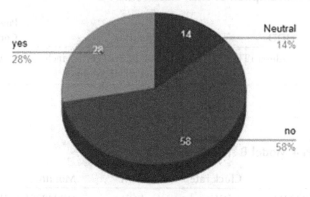

FIGURE 16.1 Opinion about the idea of replacing traditional classroom PCs with Raspberry Pi.

of Raspberry Pi 2.123 W to estimate the power consumption of the new system, which is 22.123 W. We compared the power consumption of the new system with the current system. Total energy consumption with the new system for one day would be around: 22.12×8×65 Wh=11.50 KWh. *So there is a reduction of power consumption by 44.72 /11.50 = 3.88 times.*

The drivers and barriers of replacing the traditional PCs with thin clients in the regular classroom were seen in the interviews that were taken from the people who are involved in using the computers in the classroom. The faculties are directly involved in the usage of classroom computers.

East West University has a total of 13 departments. All the faculties involved in each department directly get access to the regular classrooms for their classes. However, the usage of computers depends on the class topic and criteria of teachers' teaching processes.

Through questioning the faculties, various opinions were pointed regarding the usage of PCs and the idea of replacing them with Raspberry Pi.

It was seen that the teachers from the departments have mixed opinions about the idea of the replacement of this classroom PC. Some of them feel that using Raspberry Pi can be a hassle while taking classes as they cannot make proper use of it. From Figure 16.1 it is seen that the majority of the responses were negative, holding about 58 n opinions, regarding the changing device in the educational environment. Most of the negative responses came from the teachers of the Faculty of Sciences and Engineering. One of the faculties from the engineering department shared that while giving lectures it is necessary to go through several software packages in parallel to make the lecture more convenient to the students. Sometimes more than two tabs are needed to function during the classes which cannot be provided by these thin clients. They also shared that if the low-powered device cannot provide sufficient flexibility while browsing the lecture topics, it might cause inconvenience.

About 14% opinion went for neutral as well as mixed responses as some of the faculties were unfamiliar with the idea of thin clients. Sometimes they do not need to use PCs for taking the class. Another senior lecturer from the Faculty of Liberal Arts and Social Sciences shared that the idea of the thin clients was unfamiliar, therefore the idea of replacing the classroom PCs with these devices might take time to adopt if a replacement needs to occur. However, they appreciated the idea of the replacement as it would be energy efficient. About 28% of respondents have given positive opinions regarding this study as they believe that it would be a great idea to embrace new technologies for education purposes if they can be easily adapted and eco-friendly.

According to the findings of this study, a PC consumes 86 W of electricity per hour in one classroom, a Raspberry Pi consumes 1.42 W, and a monitor consumes 21 W. In total, 22.42 W of power is used. As a result, using Raspberry Pi will save a significant amount of electricity. As a result, it will be less expensive. However, because some engineering teachers employ heavy software, the Raspberry Pi will not be acceptable in such situations. Furthermore, because most teachers are unaware of this gadget, they will be unwilling to employ it. And, during the survey, a significant number of teachers expressed their satisfaction with the traditional PC.

University of Technology's Digital Forensic Research Labs in New Zealand. They delved into the applications of low-cost computing using the Raspberry Pi 2 Model B. Their research aimed to leverage the affordability and versatility of this single-board computer for various purposes, potentially benefiting digital forensics and expanding access to advanced computing in resource-constrained environments. The study [2] explored the energy efficiency of this popular single-board computer, offering insights into its practical implications and potential environmental impact. The study aimed to provide valuable information for optimizing power usage in Raspberry Pi-based applications. Through this study, it can be concluded that using Raspberry Pi instead of a PC in the classrooms will be highly beneficial for both power consumption and economic expenditure. But the teachers show a negative demeanor about changing the PC to something they are not used to.

16.5 LIMITATIONS AND FUTURE WORK

The data regarding the power consumption is collected from the internet. As a result, it may differ in practice. And also, there are some other constraints like using the projector, sound system and so on, that were not addressed in the study. In the future, a more precise computation will be useful, as will the addition of the other constraints.

16.6 CONCLUSION

Modern technology integration in the education industry has become essential in today's information-driven world. A cost-effective and durable computing environment must be created to raise educational standards in poor nations. In Bangladesh, practically all educational institutions use computers for all aspects of work, similar to other emerging nations. The high cost and energy usage of conventional PCs present problems, though.

Many of these tasks in the classroom can be carried out with a basic computer that consumes less energy as not all of them demand a powerful computer. Consequently, replacing these more energy-intensive equipment can have a good effect on the environment. A wonderful alternative to conventional PCs is the Raspberry Pi. It is a tiny, single-board computer that costs nothing. Due to its inexpensive price and relatively low average power consumption when compared to desktop and laptop computers, it has the potential to save money and energy.

Per day in only one classroom the amount of power consumption by a PC is 688 Wh whereas the consumption of Raspberry Pi is 176.96 Wh. So in a day, in one classroom 511.04 Wh of energy can be saved. It is fair to mention the economic benefits will be huge. Even though many constraints remain there. Raspberry Pi will save a lot of power and energy, there is no denying that. But changing a whole system can be cumbersome. And moreover getting used to a different system and environment is not always welcomed. So some teachers also show their lack of interest in this change. As change is a part of life, if the change is for the better then some compromises can be made. Similarly in this case Raspberry Pi can save a huge amount of energy which will decrease an appreciable amount of electricity usage along with environmental pollution. So such a system should be encouraged.

REFERENCES

1. Kyaw, A.K., Truong, H.P., & Joseph, J. (2018). "Low-Cost Computing Using Raspberry Pi 2 Model B." *J. Comput.*, 13(3), 287–299.
2. Bekaroo, G., & Santokhee, A. (2016, August). "Power consumption of the Raspberry Pi: A comparative analysis." In *2016 IEEE International Conference on Emerging Technologies and Innovative Business Practices for the Transformation of Societies* (EmergiTech) (pp. 361–366). IEEE.
3. Chowdhury, M.T., et al. (2023). "Comparison of feasibility between machine learning algorithms in terms of predicting energy consumption of smart grid." In *2023 4th International Conference on Big Data Analytics and Practices (IBDAP)* (pp. 1–6). doi: 10.1109/IBDAP58581.2023.10272003.
4. Hossain, M.S., et al. (2023). "Performance evaluation of intrusion detection system using machine learning and deep learning algorithms." In *2023 4th International Conference on Big Data Analytics and Practices (IBDAP)* (pp. 1–6). doi: 10.1109/IBDAP58581.2023.10271964.
5. Chowdhury, M.T., Rahman, H., Sumon, M.I., & Talha, A. (2024). "Classification of satellite images with VGG19 and Convolutional Neural Network (CNN)." In *2024 2nd International Conference on Advancement in Computation & Computer Technologies (InCACCT)* (pp. 397–402). doi: 10.1109/InCACCT61598.2024.10551185.
6. Handayani, R., Siregar, S., Ike Sari, M., & Afrizal, G. (2018). "Thin client system for education purpose using raspberry pi." *Int. J. Eng. Technol.*, 7(4.44), 233. doi:10.14419/ijet.v7i4.44.26994
7. Pattinson, C., Cross, R., & Kor, A.L. (2015). "Thin-client and energy efficiency." In *Green Information Technology* (pp. 279–294). Morgan Kaufmann.
8. Mollova, S., Simionov, R., & Seymenliyski, K. (2018, October). "A study of the energy efficiency of a computer cluster." In *Proceedings of the Seventh International Conference on Telecommunications and Remote Sensing* (pp. 51–54).

9. Kirkness, A. (2021). "Feasibility review of a start-up full-service freelance ICT firm in NZ." In *Proceedings of the IEEE Conference on Asia Entrepreneurship and Sustainability* (pp. 50–92).
10. Kyaw, A.K., Truong, H.P., & Joseph, J. (2018). "Low-cost computing using Raspberry Pi 2 Model B." In *Proceedings of IEEE Conference on Computing* (pp. 287–299).

Index

A

Access control, 232, 233
Accuracy, 19
  evaluation for bright and low brightness images, 201–203, 212
Adaptive compression algorithms, 255
Advanced Encryption Standard (AES), 235
Affordability
  benefits of Raspberry Pi for classrooms, 288, 290
Agent network, 228
Agricultural challenges
  climate change, land degradation, fluctuating food prices, 89
Agricultural Information Management Systems (AIMS), 77–78
Agricultural robots
  AI applications, 90
AI adoption in public sector
  trends, 122
AI algorithms
  use in fraud detection, 112
AI in access control
  behavioral biometrics and identity verification, 183
AI in agriculture, 70, 79
AI in anomaly detection
  applications in cybersecurity, 177
AI in civil engineering
  impacts, 115–116
AI model orchestration
  integration with XAI, 130–131
AI-powered chatbots
  enhancing citizen engagement, 117
Algorithmic humanitarianism
  refugee status determination, 112
Algorithm optimization, 26, 29, 33
Alternative computing solutions
  energy-efficient options, 290
Alzheimer's disease
  drug discovery and treatment planning, 163–164
Anomaly detection
  application in fraud prevention, 113
Applications of QML
  drug discovery, disease diagnosis, and treatment planning, 161–164
Artificial intelligence (AI)
  applications, 7–9
  applications in healthcare, 48–49
  in diagnostics, 27, 34
  ethical considerations, 64
  tools and benefits in farming, 89–91
Attendance management systems
  biometric and automated solutions, 258–260
Attendance tracking
  cloud integration and mobile applications, 259
Auto image forgery detection
  challenges and applications, 197–199
Automation in green computing
  role in IT efficiency, 285
Autonomous drones, 91
Awareness of green computing
  importance in education, 278–279, 283

B

Behavioral analysis, 233, 237, 238
  role in predictive threat intelligence, 177
Bias in data models, 31, 35
Big data analytics, 71, 84
Biomarkers
  autoimmune, 5
  identification, 3, 6
Biometric authentication
  fraud prevention and multi-factor systems, · 258, 265
Blockchain
  supply chains and frameworks, 72–74
Bright color images
  accuracy and precision results, 201–203, 210
Broad network access, 221

C

Camera requirements, 266
Carbon emissions
  impact of traditional PCs, 289
  reduction through green computing, 279–280
Cardiovascular disease, 2, 5–6
Case studies in HIS optimization
  data analytics, 65
  EHR optimization, 65
Challenges in AI/ML security systems, 240
Charge-coupled devices (CCDs), 245
Citizen engagement
  transformation with AI, 117
Classical machine learning
  comparison with QML, 165–167

299

Classroom energy consumption
  comparison between PCs and Raspberry Pi, 292–293
Clinical decision support systems, 28, 33, 38
Cloud-based services
  benefits in higher education, 282
  use in XAI deployment, 142
Cloud computing, 71, 80
Cloud computing integration, 259
Cloud environments
  AI-enhanced security strategies, 185
Clustering methods, 30, 34, 42
Cocoa industry
  disease challenges, economic significance, 101–102
Coherence in qubits
  role in quantum computing, 156
Collaborative security ecosystems, 241
Compression techniques
  lossless and lossy, 253–254
Container orchestration
  challenges in XAI deployment, 132–133
Controlled datasets
  benefits for forgery detection, 199–200
Copy-move forgery detection
  techniques and applications, 192, 198
Cost-effectiveness
  thin client implementation, 290
Crop monitoring, 80–81
CT imaging, 249
Cybersecurity
  role of AI in risk assessments, 186

D

Darktrace, AI cyber defense, 242
Data analysis, 76, 80
  power consumption comparison, 291–292
Data analytics
  decision-making insights, 49, 65
Data in transit, 232, 233
Data preprocessing, 25, 32, 38
Datasets
  controlled vs. uncontrolled, 199
Data validation techniques
  schema alignment in XAI, 145
Decision-making
  role of AI in welfare, 111
Decision tree algorithms, 225, 229
Diabetes
  prevalence, 2
  Type 1 and 2 characteristics, 4–5
Digital agriculture
  AI and IoT transformation, 91
Digital divide
  addressing with low-cost computing, 290

Digital Imaging and Communications in Medicine (DICOM), 250
Digital image processing (DIP), 248
Distributed denial of service (DDoS), 225
Drone path planning
  bio-inspired optimization algorithms, 122
Drug discovery
  QML applications in virtual screening, 162
Dynamic threat analysis
  real-time monitoring and response, 181

E

Electronic Health Records (EHR)
  integration challenges, 49, 65
Energy efficiency
  Raspberry Pi *vs.* traditional PCs, 292–293
ENERGY STAR-certified equipment
  impact on sustainability, 283
Environmental impact
  reducing with green computing, 288–289
Ethics in AI healthcare, 39, 46
E-waste
  responsible disposal practices, 281–282
Explainable AI (XAI)
  challenges in quantum systems, 169
  importance in information system security, 188
  principles and architecture, 128–129

F

Face detection
  Haar Cascade Classifier, 266–269
Face recognition
  real-time applications and benefits, 258, 262
False Positive Rate (FPR), 227
Fault injection
  testing XAI system robustness, 145
Feature engineering, 17
  feature selection, 13
Feature extraction
  SURF and SIFT integration, 192, 195
Feature selection, 29, 36, 42
Feature selection in machine learning, 61
Food security
  role of agriculture, 89
Fraud detection
  AI applications, 112
Fraud prevention in banking
  biometric data and AI, 114

G

Gaussian Naïve Bayes (GNB), 228
GDPR compliance 260
Genomic data analysis, 15
Genomic data analysis
  QVAE for compression and analysis, 160

Index

Green computing
  principles and strategies, 278–279
Green data centers
  energy efficiency in higher education, 279

H

Haar Cascade Classifier 266–269
HbA1c levels, 6
Health Information Systems (HIS)
  architecture and components, 50
  optimization principles, 54
Healthcare data integration, 31, 37, 44
Healthcare triage
  benefits of AI, 120–121

I

IBM Security QRadar SIEM, 242
Image acquisition and enhancement, 248
Image forgery detection
  automation and performance analysis, 197–199
Immigration decision-making
  ethical concerns, 112
Indexed color images, 247
Infrastructure planning
  AI in design and optimization, 115–116
Insider threats
  detection using behavioral analysis, 175
Insulin resistance, 6
Internet of Medical Things (IoMT)
  role in HIS, 67
Interpixel redundancy, 252
Interviews
  faculty perspectives on thin clients, 291
Intrusion detection systems (IDS), 220, 225–226
IoT (Internet of Things), 260, 263
  in agriculture, 72, 79
  applications in farming, 90
Istio
  proxy injection in Kubernetes for XAI, 136–137
IT sustainability plans
  implementation in universities, 283

K

Key management, 234, 236
K-Means clustering, 225, 228
Kubernetes
  managing XAI deployments, 131–132

L

LIME (Local Interpretable Model-Agnostic Explanations)
  case study, 130

Limitations
  challenges of implementing thin clients, 295
Linear regression, 226
Lossless compression, 253
Lossy compression, 254

M

Machine learning (ML)
  applications in HIS, 54
  applications in smart farming, 71, 79
  KNN *vs.* SVM comparison, 272–273
  model development, 61
  models, 26, 34, 38
Malware detection
  signature-based *vs.* behavior-based approaches, 180
MATLAB
  simulation tool description, 196
Mean Absolute Error (MAE), 226
Medical imaging
  clustering and QML applications, 162
  evolution and significance, 249
Methodology for green computing adoption, 281
Metrics for threat detection, 237
Metrics monitoring
  telemetry improvements, 133–134
Model validation, 32, 41
Modernization needs in African agriculture, 95
Multi-factor authentication (MFA)
  AI's role, 184

N

Natural language processing (NLP)
  analyzing security text data, 178
Network security, 220
Neural cryptography, 237
Neural network explanation
  integration with XAIaaS, 145
Neural networks, 28, 33, 45

O

Optimization
  quantum techniques in personalized treatment, 162–164
Optimization of farming efficiency, 90
Optimization techniques
  healthcare resource allocation, 56

P

Persistent volumes
  role in stateful services, 145
Policy development
  fostering sustainable practices, 282–283

Power consumption
  reduction with Raspberry Pi, 293–294
Precision
  comparison for bright and low brightness images, 203
Precision agriculture
  techniques and benefits, 71, 80, 84
Precision farming
  technological advancements, 95
Precision medicine, 3
Predictive analytics, 25, 31, 37
  disease management, 61
  proactive healthcare applications, 121
Predictive encryption, 237
Predictive threat intelligence
  applications in cybersecurity, 177
Prevalence
  evaluation metrics for image recognition, 209–210
Privacy concerns
  challenges in AI security, 187
Privacy concerns and GDPR compliance, 260
Procurement
  eco-conscious IT purchasing decisions, 281
Public sector integration
  AI benefits, 122

Q

QAOA (Quantum Approximation Optimization Algorithm)
  applications in molecular biology, 162
Quantum encryption, 235, 237
QVAE (Quantum Variational Autoencoder)
  biomedical data compression, 160

R

Random forest algorithms, 29, 35
Raspberry Pi
  advantages in educational settings, 290, 293
Real-time systems
  video processing, 270
Recall
  metrics for bright and low brightness images, 207–208
Region of Interest (ROI), 255
Regression analysis, 32, 40
Remote sensing applications, 252
Renewable energy sources
  integration in green IT, 280
Research ethics
  data collection procedures, 291
Resource allocation
  optimizing welfare programs, 111
Resource pooling, 221

Risk assessment
  AI-powered approaches, 186
Robotics
  seeding and pest control, 80–82
Rubber industry
  challenges and genetic improvements, 99–100

S

Schema mismatch
  impact on XAI operations, 144
Secure communication protocols, 238
Security enhancements
  multi-layered biometric systems, 265–266
Sensitivity and specificity, 11
Sensors
  integration in agriculture, 73
Service discovery
  in containerized XAI deployments, 139
SIFT algorithm
  comparison with SURF, 193–195
Simulation results
  combined SIFT+SURF model, 195
Smart agriculture
  overview and applications, 89, 91
Smart optimization techniques
  application in public services, 125
Speech recognition
  assistive technologies, 64
SQLite database, 268
Stakeholder involvement
  driving sustainability in IT, 283
Student engagement
  role in promoting green IT awareness, 279
Superposition
  concept in quantum computing, 156
Supply chain security
  AI strategies, 176
Support vector machine (SVM), 225, 229
  quantum $vs.$ classical, 159
SURF algorithm
  implementation and evaluation, 192, 194
Sustainability
  integrating green computing in classrooms, 288–289
Symptoms
  onset and reduction, 3, 6

T

Technology integration
  digital farming solutions, 103
Technology's environmental impact
  addressing challenges, 280
Telemedicine
  role in HIS, 64

Index

Telemedicine applications, 251
Telemetry
    results and incident tracking in XAI, 144
Thin client solutions
    potential benefits and barriers, 289–291
Threat detection
    AI for proactive security measures, 173
Thresholding in image processing, 254
TLS/SSL protocols, 238
Traceability and smart farming, 72
Traditional PCs
    energy-intensive challenges, 288
Training datasets, 25, 34, 38
Treatment planning
    role of QML in personalized medicine, 163
Triage systems
    predictive analytics in healthcare, 120
True and false positives
    impact on accuracy, 200–203
Trust-based models, 225

U

University initiatives
    fostering green IT practices, 282–284
Unsupervised learning, 7, 30, 36, 223
User behavior analysis (UBA), 233, 237

V

Variational Quantum Eigen Solver (VQE)
    protein structure prediction, 166
Vector quantization, 254

W

Waste reduction
    role of sustainable computing, 288
Wearable technology, 15
Weather prediction, 83
Workforce
    challenges in rubber and cocoa industries, 101

X

XAI in cybersecurity
    applications and challenges, 188
XAI in QML
    limitations and future directions, 169
XAIaaS (Explainable AI as a Service)
    overview and future work, 128, 145

Y

YOLO models
    role in recognition, 264